AF576006

Latest Advancements in Next-Generation Semiconductors: Materials and Devices for Wide Bandgap and 2D Semiconductors

Latest Advancements in Next-Generation Semiconductors: Materials and Devices for Wide Bandgap and 2D Semiconductors

Editors

Zeheng Wang
Jingkai Huang

Basel • Beijing • Wuhan • Barcelona • Belgrade • Novi Sad • Cluj • Manchester

Editors
Zeheng Wang
Data61
CSIRO
Sydney
Australia

Jingkai Huang
Department of Systems
Engineering
City University of Hong Kong
Hong Kong
China

Editorial Office
MDPI
St. Alban-Anlage 66
4052 Basel, Switzerland

This is a reprint of articles from the Special Issue published online in the open access journal *Micromachines* (ISSN 2072-666X) (available at: www.mdpi.com/journal/micromachines/special_issues/3LEPW7T95G).

For citation purposes, cite each article independently as indicated on the article page online and as indicated below:

Lastname, A.A.; Lastname, B.B. Article Title. *Journal Name* **Year**, *Volume Number*, Page Range.

ISBN 978-3-0365-9511-5 (Hbk)
ISBN 978-3-0365-9510-8 (PDF)
doi.org/10.3390/books978-3-0365-9510-8

Contents

About the Editors

Zeheng Wang

Zeheng is a CERC Fellow, appointed through the Impossible Without You campaign. He earned his PhD in Si quantum computing devices from CQC2T, UNSW, and also has significant experience in III-V semiconductor devices, semiconductor physics and modeling, machine learning, and TCAD simulations. Zeheng has published over 30 scientific papers in top-tier venues, including Nature, Advanced Materials (cover article), Energy & Environmental Science, IEEE Electron Device Letters, and IEEE Transactions on Electron Devices, as well as flagship conferences such as IEEE IEDM and IEEE ISPSD.

Zeheng's expertise is recognized by his service as an editor for *Micromachines* (IF3.5) and *Applied Research* (Wiley). He is also an active reviewer for many leading journals published by IEEE, IET, IEICE, Wiley, Springer, Elsevier, and IOP. He currently conducts more than 20 papers' peer review for SCI-indexed journals per year. His open-source online service for semiconductor fabrication has been used by over 300 researchers worldwide. Additionally, he invented two patents in the field of semiconductor devices.

Zeheng's current research interests include quantum sensing, quantum artificial intelligence, artificial intelligence for science (AI4Science, particularly in materials, electronics, and medicines), and micro/nano manufacturing and fabrication.

Jingkai Huang

Dr. Huang is a seasoned researcher with a robust background in semiconductor R&D. Prior to City University of Hong Kong, he contributed significantly to the field at King Abdullah University of Science and Technology (2015–2019, KSA), TSMC (2019, TW), and National Nano Device Laboratories (2010–2011, TW). His expertise lies in the nanofabrication of ultra-scaled electronic devices and the scalable manufacture of low-dimensional materials for advanced electronics. Outstanding achievements led Dr. Huang to be awarded the prestigious Scientia Ph.D. scholar for the "Beyond the Silicon Technology" project at UNSW. His current research focuses on integrating emerging materials such as 2D semiconductors, complex oxides (high- κ, ferroelectric, and memory), and porous coordination polymers (low-κ and sensor) in order to advance future semiconductor technology nodes.

MDPI

Editorial

Latest Advancements in Next-Generation Semiconductors: Materials and Devices for Wide Bandgap and 2D Semiconductors

Zeheng Wang [1,*] and Jing-Kai Huang [2,3,*]

[1] Manufacturing, CSIRO, Lindfield, NSW 2070, Australia
[2] Department of Systems Engineering, City University of Hong Kong, Kowloon, Hong Kong
[3] School of Materials Science and Engineering, University of New South Wales (UNSW), Sydney, NSW 2052, Australia
* Correspondence: zenwang@outlook.com (Z.W.); jkhuang@cityu.edu.hk (J.-K.H.)

Semiconductor materials, devices, and systems have become indispensable pillars supporting the modern world, deeply ingrained in various facets of our daily lives [1,2]. In the realm of computing, semiconductors serve as the backbone of microprocessors and memory units, enabling the rapid calculations and data storage that power everything from personal computers to large-scale data centers [3–5]. When it comes to power conversion, semiconductor-based components like transistors and diodes are critical in optimizing energy efficiency, whether in renewable energy systems or electric vehicles [6–9]. Moreover, in the field of optoelectronics, semiconductor technologies have revolutionized lighting and detecting solutions through the development of energy-efficient and long-lasting light-emitting diode (LED) technology [10–13], which has not only transformed the way we illuminate our surroundings, but also enhanced the precision and sensitivity of various detection and sensing applications [14–16]. Moreover, in the ever-expanding domain of information processing, semiconductors are at the heart of information storage, communication devices, and complex integrated circuits that facilitate the seamless exchange and long-duration storage of data [17,18].

The Special Issue on "Latest Advancements in Next-Generation Semiconductors: Materials and Devices for Wide Bandgap and 2D Semiconductors" serves as a comprehensive repository of cutting-edge research that extends the frontiers of semiconductor devices, optoelectronics, and material science. The articles in this issue delve into a myriad of challenges and opportunities that characterize these rapidly evolving disciplines. Specifically, the contributions are organized around four pivotal themes.

Citation: Wang, Z.; Huang, J.-K. Latest Advancements in Next-Generation Semiconductors: Materials and Devices for Wide Bandgap and 2D Semiconductors. *Micromachines* **2023**, *14*, 1992. https://doi.org/10.3390/mi14111992

Received: 23 October 2023
Accepted: 25 October 2023
Published: 27 October 2023

1. Device Architecture and Design

The field of semiconductor devices has experienced significant advancements, particularly in the design and performance of high-electron-mobility transistors (HEMTs) [19,20]. For instance, the paper by Meng Zhang et al. explores the influence of gate geometry on AlGaN/GaN nanochannel HEMTs, revealing that tri-gate designs offer higher peak transconductance and frequency performance compared to dual-gate structures. Another study by Haiwu Xie et al. delves into the irradiation effects on tunnel field-effect transistors, offering optimization strategies for better reliability against irradiation.

This Special Issue also covers innovative device designs and the challenges associated with crucial environments. A study by Choi and colleagues on AlGaN/GaN HEMTs with HfO2 as the passivation layer shows improvements in breakdown voltage, but highlights the need for balancing this with frequency characteristics. Similarly, research on quartz flexible accelerometers by Zhigang Zhang et al. discusses the suppression of noise through high-pass filtering.

2. Material Science and Fabrication Techniques

Material science and fabrication engineering play crucial roles in the performance and reliability of electronic devices [21–23]. Han et al. investigate the impact of oxygen flow rates on the electrical characteristics of amorphous indium–gallium–zinc oxide transistors, emphasizing the importance of controlling oxygen vacancies. In another study, Kudryashov and colleagues explore the use of ultrashort laser pulses for inscribing photoluminescent micro-bits inside dielectric crystals, opening new avenues for optomechanical memory storage. These two developments indicate that using proper material science and fabrication engineering innovations can still further boost the performance of semiconductor components [24,25].

3. Energy Efficiency and Power Management

Energy efficiency and power management are other areas that have received considerable attention in semiconductor-based systems. For example, a study by Weizhong Chen et al. on FIN-LDMOS with a bulk electron accumulation effect shows promising results in terms of specific on-resistance, which may be beneficial for high-power telecommunication applications [26]. Another paper by Boyang Ma et al. focuses on enhancing the ESD performance of power-rail clamp circuits, offering a design that is strengthened against false triggers [27]. In addition, power management is a critical aspect of modern electronic systems. One paper by Dan Butnicu in this Special Issue focuses on the derating-sensitive failure rate of tantalum polymer capacitors within DC-DC eGaN-FET-based PoL converters, providing insights into improving the reliability of these systems [28].

4. Optoelectronics and Light-Based Technologies

Lastly, this issue features comprehensive reviews that offer a broader perspective on this field. One such review by Zhaoyong Liu et al. focuses on the advancements in integrating meta-surface structures with micro-LEDs, providing a roadmap for future research [15].

In summary, the articles in this Special Issue collectively contribute to the advancement of micro- and nanoelectronics, offering innovative solutions and highlighting areas that require further investigation. As the guest editors, it is our privilege to present this collection of pioneering research that will undoubtedly serve as a valuable resource for scholars and industry professionals alike.

Author Contributions: Conceptualization, Z.W. and J.-K.H.; writing—original draft preparation, Z.W.; writing—review and editing, J.-K.H.; All authors have read and agreed to the published version of the manuscript.

Conflicts of Interest: The authors declare no conflict of interest.

List of Contributions:

1. Butnicu, D. A Derating-Sensitive Tantalum Polymer Capacitor's Failure Rate within a DC-DC eGaN-FET-Based PoL Converter Workbench Study. *Micromachines* **2023**, *14*, 221. https://doi.org/10.3390/mi14010221.
2. Chen, W.; Duan, Z.; Zhang, H.; Han, Z.; Wang, Z. A FIN-LDMOS with Bulk Electron Accumulation Effect. *Micromachines* **2023**, *14*, 1225. https://doi.org/10.3390/mi14061225.
3. Choi, J.-H.; Kang, W.-S.; Kim, D.; Kim, J.-H.; Lee, J.-H.; Kim, K.-Y.; Min, B.-G.; Kang, D.M.; Kim, H.-S. Enhanced Operational Characteristics Attained by Applying HfO_2 as Passivation in AlGaN/GaN High-Electron-Mobility Transistors: A Simulation Study. *Micromachines* **2023**, *14*, 1101. https://doi.org/10.3390/mi14061101.
4. Han, Y.; Lee, D.H.; Cho, E.-S.; Kwon, S.J.; Yoo, H. Argon and Oxygen Gas Flow Rate Dependency of Sputtering-Based Indium-Gallium-Zinc Oxide Thin-Film Transistors. *Micromachines* **2023**, *14*, 1394. https://doi.org/10.3390/mi14071394.
5. Kudryashov, S.; Danilov, P.; Smirnov, N.; Kuzmin, E.; Rupasov, A.; Khmelnitsky, R.; Krasin, G.; Mushkarina, I.; Gorevoy, A. Photoluminescent Microbit Inscripion Inside Dielectric Crystals

by Ultrashort Laser Pulses for Archival Applications. *Micromachines* **2023**, *14*, 1300. https://doi.org/10.3390/mi14071300.
6. Liu, Z.; Ren, K.; Dai, G.; Zhang, J. A Review on Micro-LED Display Integrating Metasurface Structures. *Micromachines* **2023**, *14*, 1354. https://doi.org/10.3390/mi14071354.
7. Ma, B.; Chen, S.; Wang, S.; Qian, L.; Han, Z.; Huang, W.; Fu, X.; Liu, H. A False Trigger-Strengthened and Area-Saving Power-Rail Clamp Circuit with High ESD Performance. *Micromachines* **2023**, *14*, 1172. https://doi.org/10.3390/mi14061172.
8. Xie, H.; Liu, H. Single-Particle Irradiation Effect and Anti-Irradiation Optimization of a JLTFET with Lightly Doped Source. *Micromachines* **2023**, *14*, 1413. https://doi.org/10.3390/mi14071413.
9. Zhang, M.; Chen, Y.; Guo, S.; Lu, H.; Zhu, Q.; Mi, M.; Wu, M.; Hou, B.; Yang, L.; Ma, X.; et al. Influence of Gate Geometry on the Characteristics of AlGaN/GaN Nanochannel HEMTs for High-Linearity Applications. *Micromachines* **2023**, *14*, 1513. https://doi.org/10.3390/mi14081513.
10. Zhang, Z.; Zhao, D.; He, H.; Tang, L.; He, Q. Analysis of Noise-Detection Characteristics of Electric Field Coupling in Quartz Flexible Accelerometer. *Micromachines* **2023**, *14*, 535. https://doi.org/10.3390/mi14030535.

References

1. Markov, I.L. Limits on Fundamental Limits to Computation. *Nature* **2014**, *512*, 147–154. [CrossRef] [PubMed]
2. Kelley, T.W.; Baude, P.F.; Gerlach, C.; Ender, D.E.; Muyres, D.; Haase, M.A.; Vogel, D.E.; Theiss, S.D. Recent Progress in Organic Electronics: Materials, Devices, and Processes. *Chem. Mater.* **2004**, *16*, 4413–4422. [CrossRef]
3. Huang, J.-K.; Wan, Y.; Shi, J.; Zhang, J.; Wang, Z.; Wang, W.; Yang, N.; Liu, Y.; Lin, C.-H.; Guan, X.; et al. High-κ Perovskite Membranes as Insulators for Two-Dimensional Transistors. *Nature* **2022**, *605*, 262–267. [CrossRef] [PubMed]
4. Huang, J.-K.; Wan, Y.; Shi, J.; Zhang, J.; Wang, Z.; Yang, Z.-L.; Huang, B.-C.; Chiu, Y.-P.; Wang, W.; Yang, N.; et al. Crystalline Complex Oxide Membrane: Sub-1 Nm CET Dielectrics for 2D Transistors. In Proceedings of the 2022 International Electron Devices Meeting (IEDM), San Francisco, CA, USA , 3–7 December 2022; IEEE: Piscataway, NJ, USA, 2022; pp. 7.6.1–7.6.4.
5. Wang, Z.; Feng, M.; Serrano, S.; Gilbert, W.; Leon, R.C.C.; Tanttu, T.; Mai, P.; Liang, D.; Huang, J.Y.; Su, Y.; et al. Jellybean Quantum Dots in Silicon for Qubit Coupling and On-Chip Quantum Chemistry. *Adv. Mater.* **2023**, *35*, 2208557. [CrossRef]
6. Wang, Z.; Cao, J.; Wang, F.; Chen, W.; Zhang, B.; Guo, S.; Yao, Y. Proposal of a Novel Enhancement Type AlGaN/GaN HEMT Using Recess-Free Field Coupled Gate. *Superlattices Microstruct.* **2018**, *122*, 343–348. [CrossRef]
7. Wang, Z.; Wang, F.; Guo, S.; Wang, Z. Simulation Study of High-reverse Blocking AlGaN/GaN Power Rectifier with an Integrated Lateral Composite Buffer Diode. *Micro Nano Lett.* **2017**, *12*, 660–663. [CrossRef]
8. Wang, F.; Chen, W.; Wang, Z.; Sun, R.; Wei, J.; Li, X.; Shi, Y.; Jin, X.; Xu, X.; Chen, N.; et al. Simulation Design of Uniform Low Turn-on Voltage and High Reverse Blocking AlGaN/GaN Power Field Effect Rectifier with Trench Heterojunction Anode. *Superlattices Microstruct.* **2017**, *105*, 132–138. [CrossRef]
9. Wang, Z.; Zhang, Z.; Wang, S.; Chen, C.; Wang, Z.; Yao, Y. Design and Optimization on a Novelhigh-Performance Ultra-Thin Barrier AlGaN/GaN Power HEMT with Local Charge Compensation Trench. *Appl. Sci. Switz.* **2019**, *9*, 3054. [CrossRef]
10. Bergh, A.A.; Dean, P.J. Light-Emitting Diodes. *Proc. IEEE* **1972**, *60*, 156–223. [CrossRef]
11. Chang, M.-H.; Das, D.; Varde, P.V.; Pecht, M. Light Emitting Diodes Reliability Review. *Microelectron. Reliab.* **2012**, *52*, 762–782. [CrossRef]
12. Thejokalyani, N.; Dhoble, S.J. Novel Approaches for Energy Efficient Solid State Lighting by RGB Organic Light Emitting Diodes—A Review. *Renew. Sustain. Energy Rev.* **2014**, *32*, 448–467. [CrossRef]
13. DenBaars, S.P.; Feezell, D.; Kelchner, K.; Pimputkar, S.; Pan, C.-C.; Yen, C.-C.; Tanaka, S.; Zhao, Y.; Pfaff, N.; Farrell, R.; et al. Development of Gallium-Nitride-Based Light-Emitting Diodes (LEDs) and Laser Diodes for Energy-Efficient Lighting and Displays. *Acta Mater.* **2013**, *61*, 945–951. [CrossRef]
14. Wang, Z.; Wang, S.; Zhang, Z.; Wang, C.; Yang, D.; Chen, X.; Wang, Z.; Cao, J.; Yao, Y. A High-Performance Tunable LED-Compatible Current Regulator Using an Integrated Voltage Nanosensor. *IEEE Trans. Electron. Devices* **2019**, *66*, 1917–1923. [CrossRef]
15. Liu, Z.; Ren, K.; Dai, G.; Zhang, J. A Review on Micro-LED Display Integrating Metasurface Structures. *Micromachines* **2023**, *14*, 1354. [CrossRef] [PubMed]
16. Dey, A. Semiconductor Metal Oxide Gas Sensors: A Review. *Mater. Sci. Eng. B* **2018**, *229*, 206–217. [CrossRef]
17. Wang, Z.; Yang, D.; Shi, J.; Yao, Y. Approaching Ultra-Low Turn-on Voltage in GaN Lateral Diode. *Semicond. Sci. Technol.* **2020**, *36*, 014003. [CrossRef]
18. Kudryashov, S.; Danilov, P.; Smirnov, N.; Kuzmin, E.; Rupasov, A.; Khmelnitsky, R.; Krasin, G.; Mushkarina, I.; Gorevoy, A. Photoluminescent Microbit Inscripion Inside Dielectric Crystals by Ultrashort Laser Pulses for Archival Applications. *Micromachines* **2023**, *14*, 1300. [CrossRef]
19. Mishra, U.K.; Parikh, P.; Wu, Y.-F. AlGaN/GaN HEMTs-an Overview of Device Operation and Applications. *Proc. IEEE* **2002**, *90*, 1022–1031. [CrossRef]
20. Rabkowski, J.; Peftitsis, D.; Nee, H. Silicon Carbide Power Transistors: A New Era in Power Electronics Is Initiated. *IEEE Ind. Electron. Mag.* **2012**, *6*, 17–26. [CrossRef]

21. Sivula, K.; van de Krol, R. Semiconducting Materials for Photoelectrochemical Energy Conversion. *Nat. Rev. Mater.* **2016**, *1*, 15010. [CrossRef]
22. Chaves, A.; Azadani, J.G.; Alsalman, H.; da Costa, D.R.; Frisenda, R.; Chaves, A.J.; Song, S.H.; Kim, Y.D.; He, D.; Zhou, J.; et al. Bandgap Engineering of Two-Dimensional Semiconductor Materials. *Npj 2D Mater. Appl.* **2020**, *4*, 1–21. [CrossRef]
23. Neumaier, D.; Pindl, S.; Lemme, M.C. Integrating Graphene into Semiconductor Fabrication Lines. *Nat. Mater.* **2019**, *18*, 525–529. [CrossRef] [PubMed]
24. Theis, T.N.; Wong, H.-S.P. The End of Moore's Law: A New Beginning for Information Technology. *Comput. Sci. Eng.* **2017**, *19*, 41–50. [CrossRef]
25. Hoefflinger, B. ITRS: The International Technology Roadmap for Semiconductors. In *Chips 2020: A Guide to the Future of Nanoelectronics;* Hoefflinger, B., Ed.; The Frontiers Collection; Springer: Berlin/Heidelberg, Germany, 2012; pp. 161–174, ISBN 978-3-642-23096-7.
26. Chen, W.; Duan, Z.; Zhang, H.; Han, Z.; Wang, Z. A FIN-LDMOS with Bulk Electron Accumulation Effect. *Micromachines* **2023**, *14*, 1225. [CrossRef]
27. Ma, B.; Chen, S.; Wang, S.; Qian, L.; Han, Z.; Huang, W.; Fu, X.; Liu, H. A False Trigger-Strengthened and Area-Saving Power-Rail Clamp Circuit with High ESD Performance. *Micromachines* **2023**, *14*, 1172. [CrossRef] [PubMed]
28. Butnicu, D. A Derating-Sensitive Tantalum Polymer Capacitor's Failure Rate within a DC-DC eGaN-FET-Based PoL Converter Workbench Study. *Micromachines* **2023**, *14*, 221. [CrossRef]

MDPI

Article

Influence of Gate Geometry on the Characteristics of AlGaN/GaN Nanochannel HEMTs for High-Linearity Applications

Meng Zhang [1], Yilin Chen [2], Siyin Guo [1], Hao Lu [1,*], Qing Zhu [1], Minhan Mi [1,*], Mei Wu [1], Bin Hou [1], Ling Yang [1], Xiaohua Ma [1] and Yue Hao [1]

[1] School of Microelectronics, Xidian University, Xi'an 710071, China
[2] Key Laboratory of Wide Band-Gap Semiconductor Materials and Devices, School of Microelectronics, Xidian University, Xi'an 710071, China
* Correspondence: luhao@xidian.edu.cn (H.L.); mhmi@xidian.edu.cn (M.M.)

Abstract: In this study, AlGaN/GaN nanochannel high-electron-mobility transistors (HEMTs) with tri-gate (TGN-devices) and dual-gate (DGN-devices) structures were fabricated and investigated. It was found that the peak value of the transconductance (G_m), current gain cut-off frequency (f_T) and power gain cut-off frequency (f_{max}) of the TGN-devices were larger than that of the DGN-devices because of the enhanced gate control from the top gate. Although the TGN-devices and DGN-devices demonstrated flattened transconductance, f_T and f_{max} profiles, the first and second transconductance derivatives of the DGN-devices were lower than those of the TGN-devices, implying an improvement in linearity. With the nanochannel width decreased, the peak value of the transconductance and the first and second transconductance derivatives increased, implying the predominant influence of sidewall gate capacitance on the transconductance and linearity. The comparison of gate capacitance for the TGN-devices and DGN-devices revealed that the gate capacitance of the tri-gate structure was not simply a linear superposition of the top planar gate capacitance and sidewall gate capacitance of the dual-gate structure, which could be attributed to the difference in the depletion region shape for tri-gate and dual-gate structures.

Keywords: GaN; high electron mobility transistors; nanochannel; tri-gate; dual-gate

Citation: Zhang, M.; Chen, Y.; Guo, S.; Lu, H.; Zhu, Q.; Mi, M.; Wu, M.; Hou, B.; Yang, L.; Ma, X.; et al. Influence of Gate Geometry on the Characteristics of AlGaN/GaN Nanochannel HEMTs for High-Linearity Applications. *Micromachines* **2023**, *14*, 1513. https://doi.org/10.3390/mi14081513

Academic Editors: Zeheng Wang and Jingkai Huang

Received: 29 June 2023
Revised: 24 July 2023
Accepted: 25 July 2023
Published: 28 July 2023

1. Introduction

Gallium nitride (GaN)-based high-electron-mobility transistors (HEMTs) have great potential for high-frequency and high-power applications because of the advantages of heterojunction materials, including their high breakdown electric field, high two-dimensional electron gas (2DEG) sheet density and high electron mobility [1–4], leading to potential RF applications, including remote sensing, radar and wireless communication [5,6]. Except for the superiority of GaN-based HEMTs mentioned above, linearity is also a key characteristic for RF applications, especially for wireless communication. However, the transconductance nonlinearity, which is defined as the reduction of transconductance and f_T at a high drain current density level in GaN-based HEMTs, can limit the device linearity, leading to distortion of the signal [7,8]. Several structures can be used to alleviate this, such as graded polarization field effect transistors [9], double-channel heterojunctions [10], coupling-channel structures [11], N-polar HEMTs [12] and nanochannel structures [13]. Aside from these structures, nanochannel structures have attracted more attention because of the additional improvement in the electron velocity [14], self-heating effect [15], subthreshold swing [16], breakdown voltage [17] and so on. Moreover, a nanochannel structure can modulate the threshold voltage by varying the nanochannel width [18], which is also helpful for the improvement in linearity via threshold voltage synthesis and transconductance compensation [8,19,20].

For a nanochannel structure, there are two different gate structures, namely, a tri-gate structure [21,22] and dual-gate structure [23–26], which demonstrate the potential for the improvement in device linearity. Compared with tri-gate GaN HEMTs with multiple channels [27], dual-gate GaN HEMTs for multi-channel epitaxial design can achieve flattened transconductance, and thus, improve the device linearity and saturated current density simultaneously [25,26]. However, the characteristic difference between tri-gate and dual-gate structures has rarely been compared simultaneously. Moreover, the influence of source resistance nonlinearity on the linearity of GaN-based nanochannel HEMTs with tri-gate and dual-gate structures has been developed, but the influence of the capacitance from the sidewall gate was less mentioned.

In this work, AlGaN/GaN nanochannel HEMTs with tri-gate (TGN-devices) and dual-gate structures (DGN-devices) were fabricated and investigated. The DC characteristics and small-signal characteristics were compared and analyzed. The TGN-devices demonstrated a higher peak value of transconductance and cut-off frequency than that of the DGN-devices, but the DGN-devices presented lower second transconductance derivatives, implying better linearity. The gate capacitances of the TGN-devices and DGN-devices were compared and the influence of the capacitance from the sidewall gate is discussed.

2. Device Fabrication

The schematics of the TGN-devices and DGN-devices investigated in this manuscript are shown in Figure 1. The TGN-devices and DGN-devices were fabricated on the same wafer. The epitaxial heterostructure consisted of a 100 nm AlN nuclear layer, a 2 μm GaN buffer layer, a 1 nm AlN interlayer and a 20 nm AlGaN barrier layer with an aluminum composition of 23% from bottom to top, grown on a sapphire substrate. A sheet density of 369 Ω/□, a two-dimensional electron gas (2DEG) of 9.1×10^{12} cm^{-2} and 2DEG mobility of 1860 cm^2/V·s was obtained via Hall measurements at room temperature. The fabrication process of devices started with the formation of an ohmic contact on the source and drain using conventional Ti/Al/Ni/Au (20/160/55/45 nm) metal stack evaporation, followed by rapid annealing at 850 °C for 50 s in ambient N_2. After the device's electrical isolation was realized via nitrogen implantation, an ohmic contact resistance of 0.5 Ω·mm was verified using a transmission line measurement (TLM). A 120 nm SiN layer was deposited for surface passivation via plasma-enhanced chemical vapor deposition (PECVD). For the TGN-devices, as shown in Figure 1a,c, the nanochannel was surrounded and contacted from three directions by the Ni/Au gate metal, including the top and two sidewalls of the nanochannel. Figure 1e shows the cross-section FIB-SEM photo of a TGN-device. Figure 1g shows the fabrication process flow of a TGN-device. The gate foot defined a gate length (L_g) of 0.2 μm using electron beam lithography (EBL) and CF_4-based inductively coupled plasma (ICP) etching to remove SiN on the gate region. Then, a nanochannel with a width (W_{fin}) of 150, 200 or 250 nm was defined using EBL, followed by BCl_3/Cl_2-based ICP etching. As shown in Figure 1b,d, for the DGN-devices, although the nanochannel was surrounded from three directions by the Ni/Au gate metal, the nanochannel was contacted by the Ni/Au gate on the two sidewalls and there was a 120 nm thick SiN passivation dielectric between the top of the nanochannel and the Ni/Au gate. The PECVD SiN was amorphous and would not exert an influence on the polarization charges. Figure 1f shows the cross-section FIB-SEM photo of a DGN-device. Figure 1h shows the fabrication process flow of a DGN-device. The nanochannel width of the DGN-devices was equal to that of the TGN-devices and was directly realized via EBL, CF_4-based ICP etching and BCl_3/Cl_2-based ICP etching in sequence. The length of the nanochannel was equal to the gate length for both structures. The nanochannel width (W_{fin}) and the trench (W_{trench}) region were equal for both structures. The gate electrode (Ni/Au) with a gate cap length of 1.0 μm was formed via physical vapor deposition and a lift-off process. Finally, the interconnection via Ti/Au metalization for the device test was achieved. In this study, all devices had the same gate width of 100 μm, source–drain distance of 4 μm and source–gate distance of 0.9 μm.

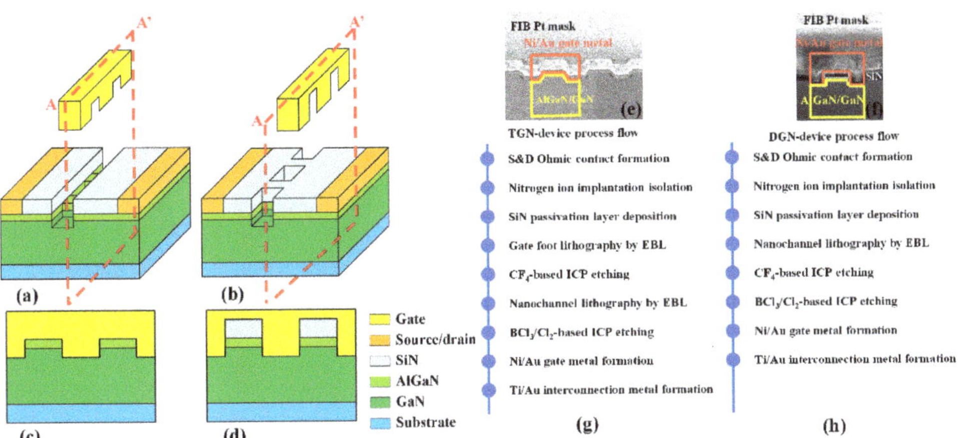

Figure 1. Schematic of a (**a**) TGN-device and (**b**) DGN-device; cross-section view of a (**c**) TGN-device and (**d**) DGN-device along the AA′ position; cross-section FIB-SEM photo of a (**e**) TGN-device and (**f**) DGN-device along the AA′ position; fabrication process flow of a (**g**) TGN-device and (**h**) DGN-device.

3. Results and Discussion

The transfer characteristics of the TGN-devices and DGN-devices are shown in Figure 2. The drain was biased at 10 V. The nanochannel widths were 150, 200 and 250 nm. The DC performance was normalized to the actual gate width. As presented in Figure 2, with the reduction in the nanochannel width, the threshold voltage (V_{th}) and the peak value of transconductance ($G_{m,peak}$) for both the TGN-devices and DGN-devices increased, which was mainly attributed to the enhanced electrostatic gate control from the sidewall gate and the reduction in the polarization charge density induced by the tensile strain relaxation [13,28]. Both the TGN-devices and DGN-devices demonstrated flattened transconductance. Figure 3 shows the V_{th} and $G_{m,peak}$ dependence on the nanochannel width (W_{fin}) for the TGN-devices and DGN-devices. As shown in Figure 3a, the gradients of the V_{th}–W_{fin} curves for the TGN-devices and DGN-devices were different. As W_{fin} increased, the V_{th} of the TGN-devices decreased slowly, while that of the DGN-devices decreased rapidly. For the TGN-devices, it was implied that the influence of the sidewall gates on V_{th} was ancillary and that of the top gate was predominant, contributing to the reduced slope of the V_{th}–W_{fin} relationship. As W_{fin} increased continuously, the V_{th} of the TGN-devices approximated the V_{th} of conventional planar GaN-based HEMTs and was limited [18]. However, for the DGN-devices, the influence of the sidewall gates on V_{th} dominated, indicating the strong dependence of V_{th} on W_{fin}. Moreover, because of the lack of control from the top gate in the dual-gate structure, as W_{fin} increased continuously, the threshold voltage of the DGN-devices decreased continuously without restriction. When further decreasing W_{fin} to less than 100 nm, the V_{th} of the TGN-devices and DGN-devices could be approximated [29], and the normally off TGN-devices and DGN devices could be realized. As shown in Figure 3b, the gradients of the $G_{m,peak}$–W_{fin} curves for the tri-gate structure and dual-gate structure were similar, indicating the similar electrostatic charge control effect from the sidewall gates. For the TGN-devices and DGN-devices with the same W_{fin}, the $G_{m,peak}$ of the TGN-devices was approximately 110 mS/mm larger than that of the DGN-devices due to the increase in the capacitance from the top gate in the tri-gate structure.

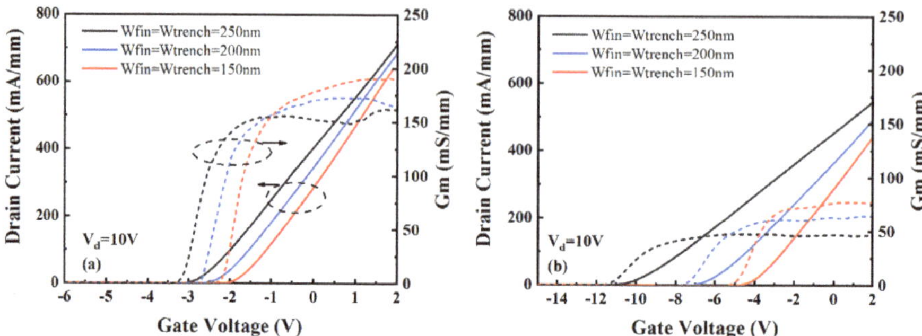

Figure 2. The transfer characteristics of the (**a**) TGN-devices and (**b**) DGN-devices. The nanochannel widths were 150, 200 and 250 nm, respectively. The straight and dotted lines standed for drain current density and tranconductance, respectively.

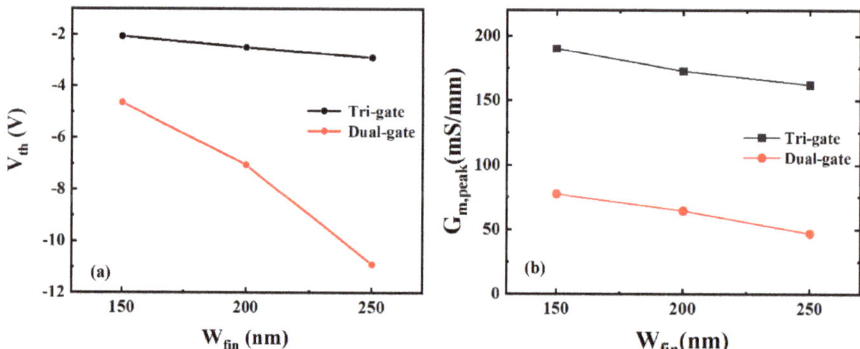

Figure 3. The dependence of (**a**) the threshold voltage (V_{th}) and (**b**) peak transconductance ($G_{m,peak}$) dependence on the nanochannel width (W_{fin}) for the TGN-devices and DGN-devices.

Figure 4 presents the gate currents for the TGN-devices and DGN-devices. As can be seen in Figure 4, for the TGN-devices, with the increase in W_{fin}, the reverse gate leakage current increased. However, for the dual-gate structure, with the increase in W_{fin}, the reverse gate leakage current was reduced. This could be attributed to the different electric field distribution dependences on W_{fin} for the tri-gate and dual-gate structures. The simulated transverse distributions of the electric field and the electric field distribution (parallel to the nanochannel length direction) of the TGN-devices and DGN-devices with W_{fin} values of 150, 200 and 250 nm are shown in Figure 5, where the simulation was performed using Silvaco Atlas [30]. The related material parameters for the simulation are listed in Table 1 [31,32]. The work function of the Schottky gate contact was set as the work function of the Ni metal (5.15 eV) [33]. Donor-type surface traps were set at the AlGaN/passivation layer interface with an activation energy of E_C-0.68 eV [34] and a constant concentration of 1.2×10^{13} cm^{-2}. These surface traps were set to compensate for the hole density on the surface [35]. The C-related traps [36] were set in the GaN buffer layer with the energy level of E_V + 0.9 eV as the deep acceptor trap and a constant concentration of 5×10^{17} cm^{-3}. The buffer traps were set to compensate for the background electron density in the GaN buffer. As shown in Figure 5a, the peak electric field in the TGN-devices occurred in the region where the top gate and heterojunction interface were in contact, as depicted in the region, implying that gate leakage primarily occurred between the barrier and the top gate. However, the peak electric field in the DGN-devices occurred in the region where the sidewall gate and heterojunction interface were in contact, as depicted in region B, implying that the gate leakage primarily occurred in the sidewall depletion region.

Figure 5b shows the peak electric field distribution along the nanochannel length direction. As shown in Figure 5b, with the increase in W_{fin}, the peak electric field increased for both the TGN-devices and DGN-devices. For the TGN-devices, the gate leakage primarily occurred in region A, which was similar to conventional planar devices. The leakage mechanism for the TGN-devices was mainly attributed to Poole–Frenkel (PF) emission and Fowler–Nordheim (FN) tunneling and the influence of the electric field was more important. Therefore, the reverse gate leakage current of the TGN-devices increased with the increase in W_{fin}. On the other hand, for the DGN-devices, the gate leakage primarily occurred in region B. During the formation of the nanochannel, the etching process could introduce etching damage and defects. The leakage mechanism for the DGN-devices was primarily associated with the sidewall-related defects, and the magnitude of leakage was jointly influenced by the electric field, trap energy levels and temperature. Moreover, the DGN-devices with smaller W_{fin} values had more sidewalls, potentially resulting in a higher total leakage current. Therefore, although the DGN-device with a W_{fin} of 150 nm presented a comparatively smaller electric field, there was still a higher total leakage current, as shown in Figure 4b.

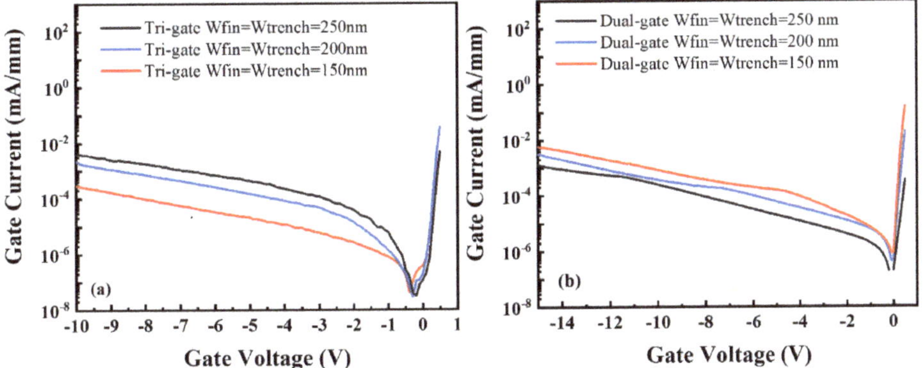

Figure 4. Gate current versus gate voltage for the (**a**) TGN-devices and (**b**) DGN-devices.

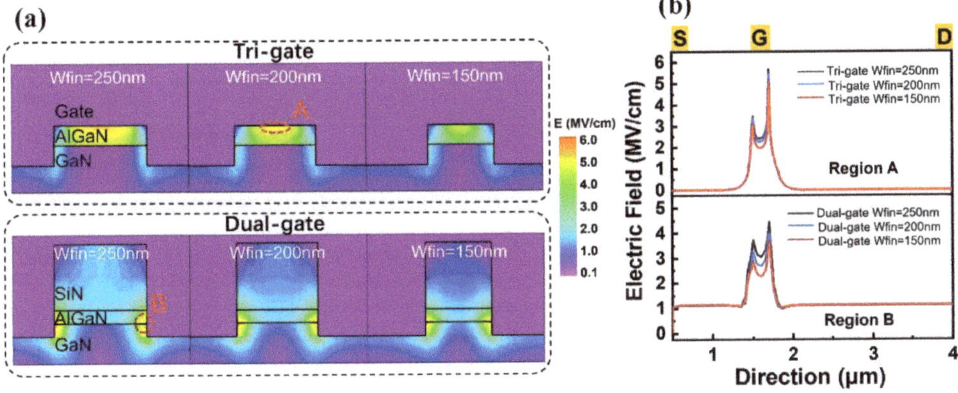

Figure 5. (**a**) The transverse distribution of the electric field and (**b**) the electric field distribution along the nanochannel length direction of the TGN-devices and DGN-devices with W_{fin} values of 150, 200 and 250 nm.

Table 1. Material parameters (bandgap, electron effective masses in the growth direction and perpendicular to the growth direction, polarization constants and lattice constants) used for the simulations [31,32]. The description of Table 1 gives the explanation of the simulation setup parameters. $m\|$ * and $m^{\perp}$ * stand for electron effective masses in the growth direction and perpendicular to the growth direction, respectively.

	GaN	AlN	$Al_{0.23}Ga_{0.77}N$
E_g (300 K) (eV)	3.42	6.28	4.08
$m\|$ *	0.18	0.25	0.20
$m^{\perp}$ *	0.20	0.33	0.23
e_{33} (C/m^2)	0.73	1.46	0.90
e_{31} (C/m^2)	−0.49	−0.60	−0.51
a_0 (Å)	3.189	3.112	3.171
c_0 (Å)	5.185	4.982	5.138

Figure 6 shows the output characteristics of the TGN-devices and DGN-devices. As shown in Figure 6, for the same W_{fin}, the TGN-devices demonstrated a higher drain current density (at gate bias of 1 V) than the DGN-devices. Moreover, for both the TGN-devices and DGN-devices, with the increase in W_{fin}, the drain current density (at a gate bias of 1V) increased, which could be attributed to the increase in the overdrive voltage. Although the gate overdrive voltage (V_g-V_{th}) of the DGN-devices was higher than that of the TGN-devices for a W_{fin} of 150 nm, the saturation current of the TGN-devices was higher than that of the DGN-devices because of the higher transconductance of the TGN-devices, which was mainly attributed to the higher product of electron mobility and gate capacitance.

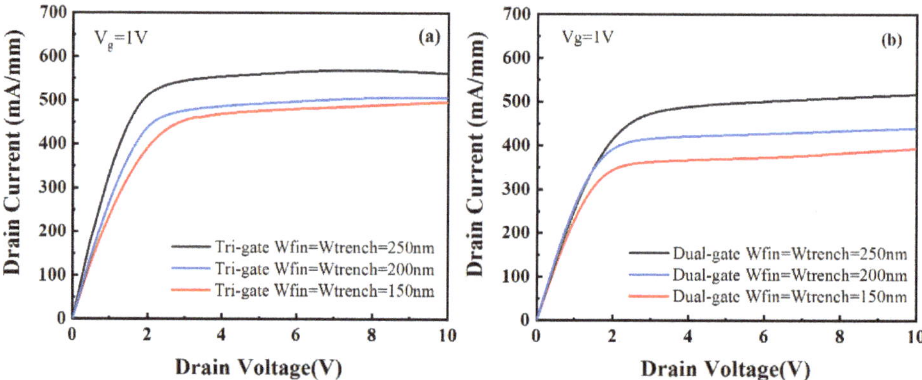

Figure 6. The output characteristics of the (**a**) TGN-devices and (**b**) DGN-devices.

Figure 7 shows the first and second transconductance derivatives (g_m', g_m'') of the TGN-devices and DGN-devices. As shown in Figure 7, for the tri-gate and dual-gate structures, as W_{fin} increased, the peak value of the first and second transconductance derivatives reduced, implying an improvement in the linearity [37]. For the same W_{fin}, compared with the TGN-devices, the DGN-devices demonstrated lower first and second transconductance derivatives, indicating better linearity characteristics. It was implied that the electrostatic control from the sidewall gates was responsible for the improvement in the linearity characteristics. Moreover, the reason for the improved linearity characteristics of DGN-devices might be that the sidewall gate electrostatic control was predominant for the dual-gate structure.

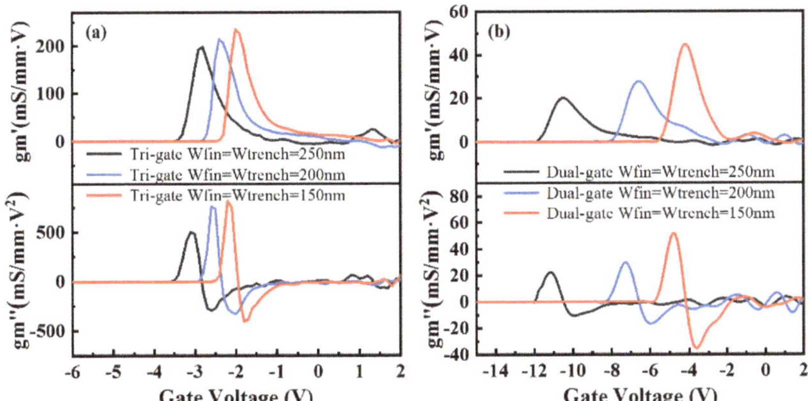

Figure 7. The first and second transconductance derivatives ($g_m{}'$, $g_m{}''$) of the (**a**) TGN-devices and (**b**) DGN-devices.

To evaluate the RF characteristics of the TGN-devices and DGN-devices, the S parameters of the devices were measured using an Agilent 8363B network analyzer within the frequency range from 100 MHz to 40 GHz. The small signal characteristics of the TGN-devices and DGN-devices (W_{fin} = 200 nm) are presented in Figure 8a. f_T and f_{max} were extracted as the intercept of the −20 dB/decade slope for H_{21} and maximum stable gain, respectively. The f_T/f_{max} of the TGN-devices and DGN-devices were 36.9/87.1 and 18/40.6 GHz, respectively. It can be seen that the TGN-devices had an about 51% higher f_T and an about 53% higher f_{max} than the DGN-devices, which was mainly attributed to the higher extrinsic transconductance of the TGN-devices. Figure 8b shows the f_T/f_{max} dependency on the gate voltage for the TGN-devices and DGN-devices with a W_{fin} of 200 nm. It is demonstrated in Figure 8b that both the TGN-devices and DGN-devices showed a flattened f_T/f_{max} versus gate voltage curve, which was attributed to the flattened transconductance curve profile for the TGN-devices and DGN-devices. Moreover, compared with the DGN-devices, the TGN-devices realized the higher f_T/f_{max} but lower gate swing of f_T/f_{max} because of the higher transconductance and the lower gate swing of transconductance for the tri-gate structure.

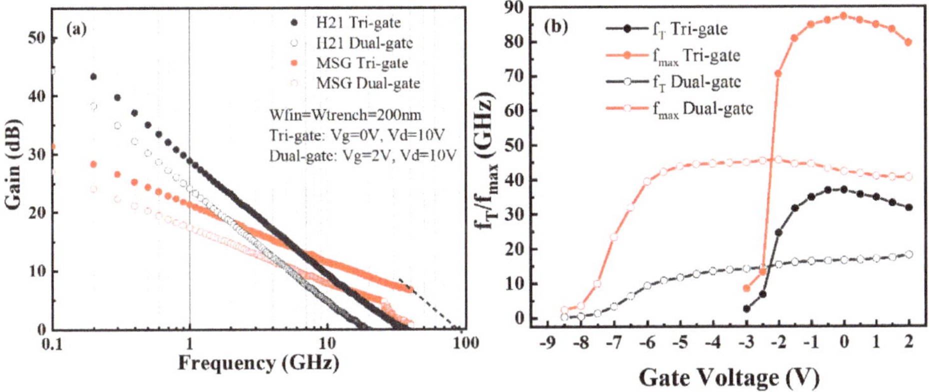

Figure 8. (**a**) Small-signal characteristics and (**b**) f_T/f_{max} as a function of the gate voltage for the TGN-devices and DGN-devices. The nanochannel width was 200 nm.

The capacitance–voltage (C-V) characteristics of the TGN-devices and DGN-devices with W_{fin} of 200 nm based on FATFET structure are shown in Figure 9. The electron sheet density (n_s) is the integral of the C-V curve:

$$n_s = \frac{1}{e}\int_{V_{pinch}}^{V_g} CdV \tag{1}$$

where e is the unit electron charge, C is the capacitance of the device, V_g is the gate voltage and V_{pinch} is the threshold voltage of the C-V curve (defined as the critical gate voltage in the C-V curve with a capacitance value less than 10 nF/cm^2). As shown in Figure 9, the gate capacitance of the TGN-devices and DGN-devices increased when the gate voltage was larger than the threshold voltage, implying that n_s nonlinearly increased with the linear increase in the gate voltage.

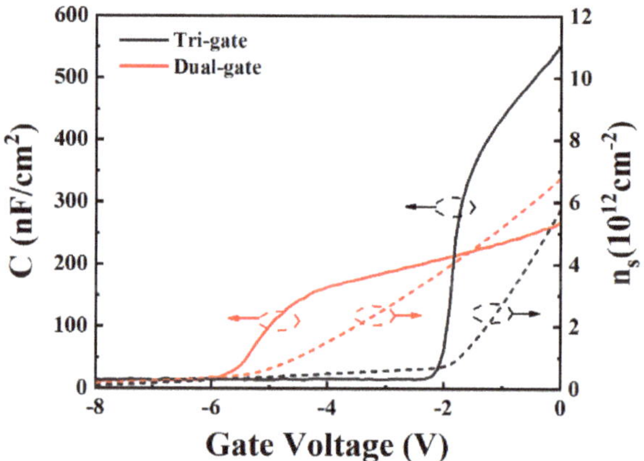

Figure 9. C-V curves and the electron sheet densities derived from the C-V curves for the TGN-devices and DGN-devices (W_{fin} = W_{trench} = 200 nm). The test structures were FATFETs with a gate width of 100 μm and gate length of 20 μm in order to minimize the parasitic component.

However, as shown in Figure 9, when the gate voltage was higher than the threshold voltage, the slope of the C-V curve for the DGN-devices was less than that of the TGN-devices, which indicated that the capacitance control of the tri-gate structure was not simply a linear superposition of top planar gate and sidewall gates (otherwise the slope of the C-V curve for both the TGN-devices and DGN-devices would be the same). Moreover, when the gate voltage was higher than the threshold voltage, the gate capacitance of the DGN-devices was lower than that of the TGN-devices, leading to the relatively slower increase in n_s, which was more similar to the MESFET-like electron channel and was attributed to improving the linearity of the devices.

The electron concentration distributions of the TGN-devices and DGN-devices for different gate overdrive voltages are shown in Figure 10. As shown in Figure 10, the TGN-devices and DGN-devices demonstrated different gate control abilities. For the TGN-devices, applying a negative voltage to the gate electrode led to the depletion of charge in both the barrier and the channel. The top gate played a significant role in this depletion process, while the sidewall gates assisted in controlling the channel near the sidewalls. As the V_g gradually became more positive, the depletion effect of the gate weakened, resulting in an increasing electron concentration in the barrier and channel. It was implied that the gate control from the top gate was dominant and the control of the electrons from the sidewall gate was assisted. For the DGN-devices, the presence of a 120 nm SiN layer between the top gate and the barrier limited the top gate's control capability over the channel but may improve the device reliability. Therefore, the control of the electrons from the sidewall gate became primary and its control effectiveness for the central region of the barrier was weaker. As the gate overdrive voltage increased, a significant difference in the electron concentration arose between the central region of the barrier and the region near the sidewall gates. Furthermore, as shown in Figure 10, under the same gate overdrive voltage, the depletion region shape for the TGN-devices and DGN-devices were different, which was attributed to the different differential C-V curves of the tri-gate and dual-gate structures. Moreover, when the gate overdrive voltage increased, as shown in Figure 10, the n_s of the TGN-devices was more than that of the DGN-devices.

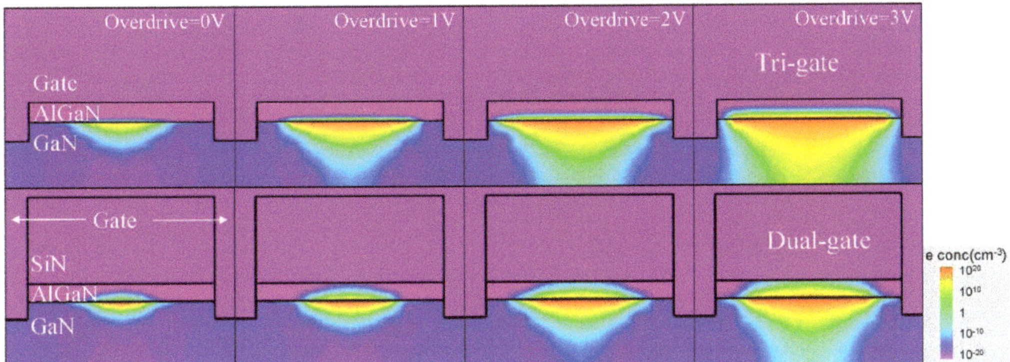

Figure 10. The electron concentration distribution along the gate width direction for the TGN-devices and DGN-devices with different gate overdrive voltages.

4. Conclusions

In this study, AlGaN/GaN nanochannel HEMTs with tri-gate (TGN-devices) and dual-gate (DGN-devices) structures were fabricated and investigated. It was found that both the TGN-devices and DGN-devices demonstrated a flattened transconductance, but the $G_{m,peak}$, f_T and f_{max} values of the TGN-devices were more than those of the DGN-devices because of the enhanced gate control ability from the top gate. For the same nanochannel width, the DGN-devices demonstrated lower second transconductance derivatives, implying better linearity characteristics compared with the TGN-devices. With the decrease in the nanochannel width, for both the TGN-devices and DGN-devices, the peak value of transconductance and the first and second transconductance derivatives increased, implying the predominant influence of sidewall gate capacitance on the transconductance and linearity. It was demonstrated that the gate capacitance of the tri-gate structure was not simply a linear superposition of the top planar gate capacitance and sidewall gate capacitance of the dual-gate structure, which could be attributed to the difference in the depletion region shape for the tri-gate and dual-gate structures.

Author Contributions: Conceptualization, M.Z. and H.L.; methodology, Y.C.; software, S.G.; validation, M.Z., H.L. and Y.C.; formal analysis, M.W.; investigation, Q.Z.; resources, B.H.; data curation, H.L.; writing—original draft preparation, M.Z.; writing—review and editing, H.L.; visualization, M.M.; supervision, L.Y.; project administration, X.M.; funding acquisition, Y.H. All authors have read and agreed to the published version of the manuscript.

Funding: This work was supported in part by the National Natural Science Foundation of China under grants 62234009, 62090014, 62188102, 62104179, 62104178 and 62104184; in part by the China National Postdoctoral Program for Innovative Talents under grant BX20200262; in part by the China Postdoctoral Science Foundation under grants 2021M692499, 2023M732730 and 2022T150505; and in part by the Fundamental Research Funds for the Central Universities of China under grant XJSJ23056.

Conflicts of Interest: The authors declare no conflict of interest.

References

1. Mishra, U.K.; Likun, S.; Kazior, T.E.; Wu, Y.-F. GaN-Based RF Power Devices and Amplifiers. *Proc. IEEE* **2008**, *96*, 287–305. [CrossRef]
2. Shinohara, K.; Regan, D.C.; Tang, Y.; Corrion, A.L.; Brown, D.F.; Wong, J.C.; Robinson, J.F.; Fung, H.H.; Schmitz, A.; Oh, T.C.; et al. Scaling of GaN HEMTs and Schottky Diodes for Submillimeter-Wave MMIC Applications. *IEEE Trans. Electron Devices* **2013**, *60*, 2982–2996. [CrossRef]
3. Wu, M.; Zhang, M.; Yang, L.; Hou, B.; Yu, Q.; Li, S.; Shi, C.; Zhao, W.; Lu, H.; Chen, W.; et al. First Demonstration of State-of-the-art GaN HEMTs for Power and RF Applications on A Unified Platform with Free-standing GaN Substrate and Fe/C Co-doped Buffer. In Proceedings of the 2022 International Electron Devices Meeting (IEDM), San Francisco, CA, USA, 3–7 December 2022; pp. 11.3.1–11.3.4. [CrossRef]
4. Hao, Y.; Yang, L.; Ma, X.; Ma, J.; Cao, M.; Pan, C.; Wang, C.; Zhang, J. High-Performance Microwave Gate-Recessed AlGaN/AlN/GaN MOS-HEMT with 73% Power-Added Efficiency. *IEEE Electron Device Lett.* **2011**, *32*, 626–628. [CrossRef]
5. Schuh, P.; Sledzik, H.; Reber, R.; Fleckenstein, A.; Leberer, R.; Oppermann, M.; Quay, R.; van Raay, F.; Seelmann-Eggebert, M.; Kiefer, R.; et al. GaN MMIC based T/R-Module Front-End for X-Band Applications. In Proceedings of the EMICC, Amsterdam, The Netherlands, 27–28 October 2008; pp. 274–277. [CrossRef]
6. Lu, H.; Hou, B.; Yang, L.; Zhang, M.; Deng, L.; Wu, M.; Si, Z.; Huang, S.; Ma, X.; Hao, Y. High RF Performance GaN-on-Si HEMTs with Passivation Implanted Termination. *IEEE Electron. Device Lett.* **2021**, *43*, 188–191. [CrossRef]
7. Palacios, T.; Rajan, S.; Chakraborty, A.; Heikman, S.; Keller, S.; DenBaars, S.; Mishra, U. Influence of the dynamic access resistance in the *gm* and *fT* linearity of AlGaN/GaN HEMTs. *IEEE Trans. Electron. Devices* **2005**, *52*, 2117–2123. [CrossRef]
8. Joglekar, S.; Radhakrishna, U.; Piedra, D.; Antoniadis, D.; Palacios, T. Large signal linearity enhancement of AlGaN/GaN high electron mobility transistors by device-level Vt engineering for transconductance compensation. In Proceedings of the IEDM, San Francisco, CA, USA, 2–6 December 2017; pp. 25.3.1–25.3.4. [CrossRef]
9. Sohel, S.H.; Rahman, M.W.; Xie, A.; Beam, E.; Cui, Y.; Kruzich, M.; Xue, H.; Razzak, T.; Bajaj, S.; Cao, Y.; et al. Linearity Improvement with AlGaN Polarization-Graded Field Effect Transistors with Low Pressure Chemical Vapor Deposition Grown SiN_x Passivation. *IEEE Electron. Device Lett.* **2019**, *41*, 19–22. [CrossRef]
10. Yu, Q.; Shi, C.; Yang, L.; Lu, H.; Zhang, M.; Wu, M.; Hou, B.; Jia, F.; Guo, F.; Ma, X.; et al. High Current and Linearity AlGaN/GaN/-Graded-AlGaN:Si-doped/GaN Heterostructure for Low Voltage Power Amplifier Application. *IEEE Electron. Device Lett.* **2023**, *44*, 582–585. [CrossRef]
11. Lu, H.; Hou, B.; Yang, L.; Niu, X.; Si, Z.; Zhang, M.; Wu, M.; Mi, M.; Zhu, Q.; Cheng, K.; et al. AlN/GaN/InGaN Coupling-Channel HEMTs for Improved gm and Gain Linearity. *IEEE Trans. Electron. Devices* **2021**, *68*, 3308–3313. [CrossRef]
12. Shrestha, P.; Guidry, M.; Romanczyk, B.; Hatui, N.; Wurm, C.; Krishna, A.; Pasayat, S.S.; Karnaty, R.R.; Keller, S.; Buckwalter, J.F.; et al. High Linearity and High Gain Performance of N-Polar GaN MIS-HEMT at 30 GHz. *IEEE Electron. Device Lett.* **2020**, *41*, 681–684. [CrossRef]
13. Zhang, M.; Ma, X.-H.; Yang, L.; Mi, M.; Hou, B.; He, Y.; Wu, S.; Lu, Y.; Zhang, H.-S.; Zhu, Q.; et al. Influence of Fin Configuration on the Characteristics of AlGaN/GaN Fin-HEMTs. *IEEE Trans. Electron. Devices* **2018**, *65*, 1745–1752. [CrossRef]
14. Arulkumaran, S.; Ng, G.I.; Kumar, C.M.M.; Ranjan, K.; Teo, K.L.; Shoron, O.F.; Rajan, S.; Bin Dolmanan, S.; Tripathy, S. Electron velocity of 6×10^7 cm/s at 300 K in stress engineered InAlN/GaN nano-channel high-electron-mobility transistors. *Appl. Phys. Lett.* **2015**, *106*, 053502. [CrossRef]
15. Wu, M.; Ma, X.H.; Yang, L.; Zhang, M.; Zhu, Q.; Zhang, X.C.; Hao, Y. Investigation of the nanochannel geometry modulation on self-heating in AlGaN/GaN Fin-HEMTs on Si. *Appl. Phys. Lett.* **2019**, *115*, 083505. [CrossRef]
16. Dai, Q.; Son, D.-H.; Yoon, Y.-J.; Kim, J.-G.; Jin, X.; Kang, I.-M.; Kim, D.-H.; Xu, Y.; Cristoloveanu, S.; Lee, J.-H. Deep Sub-60 mV/decade Subthreshold Swing in AlGaN/GaN FinMISHFETs with M-Plane Sidewall Channel. *IEEE Trans. Electron. Devices* **2019**, *66*, 1699–1703. [CrossRef]
17. Ma, J.; Erine, C.; Xiang, P.; Cheng, K.; Matioli, E. Multi-channel tri-gate normally-on/off AlGaN/GaN MOSHEMTs on Si substrate with high breakdown voltage and low ON-resistance. *Appl. Phys. Lett.* **2018**, *113*, 242102. [CrossRef]
18. He, Y.; Zhai, S.; Mi, M.; Zhou, X.; Zheng, X.; Zhang, M.; Yang, L.; Wang, C.; Ma, X.; Hao, Y. A physics-based threshold voltage model of AlGaN/GaN nanowire channel high electron mobility transistor. *Phys. Status Solidi* **2016**, *214*, 1600504. [CrossRef]
19. Choi, W.; Balasubramanian, V.; Asbeck, P.M.; Dayeh, S.A. Linearity by Synthesis: An Intrinsically Linear AlGaN/GaN-on-Si Transistor with OIP3/(F-1)PDC of 10.1 at 30 GHz. In Proceedings of the 2020 Device Research Conference (DRC), Columbus, OH, USA, 21–24 June 2020; pp. 1–2. [CrossRef]
20. Mi, M.; Wu, S.; Zhang, M.; Yang, L.; Hou, B.; Zhao, Z.; Guo, L.; Zheng, X.; Ma, X.-H.; Hao, Y. Improving the transconductance flatness of InAlN/GaN HEMT by modulating V_T along the gate width. *Appl. Phys. Express* **2019**, *12*, 114001. [CrossRef]
21. Zhang, K.; Kong, Y.; Zhu, G.; Zhou, J.; Yu, X.; Kong, C.; Li, Z.; Chen, T. High-Linearity AlGaN/GaN FinFETs for Microwave Power Applications. *IEEE Electron. Device Lett.* **2017**, *38*, 615–618. [CrossRef]
22. Zheng, Z.; Song, W.; Lei, J.; Qian, Q.; Wei, J.; Hua, M.; Yang, S.; Zhang, L.; Chen, K.J. GaN HEMT with Convergent Channel for Low Intrinsic Knee Voltage. *IEEE Electron. Device Lett.* **2020**, *41*, 1304–1307. [CrossRef]
23. Shinohara, K.; King, C.; Carter, A.D.; Regan, E.J.; Arias, A.; Bergman, J.; Urteaga, M.; Brar, B. GaN-Based Field-Effect Transistors with Laterally Gated Two-Dimensional Electron Gas. *IEEE Electron. Device Lett.* **2018**, *39*, 417–420. [CrossRef]

24. Odabasi, O.; Yilmaz, D.; Aras, E.; Asan, K.E.; Zafar, S.; Akoglu, B.C.; Butun, B.; Ozbay, E. AlGaN/GaN-Based Laterally Gated High-Electron-Mobility Transistors with Optimized Linearity. *IEEE Trans. Electron. Devices* **2021**, *68*, 1016–1023. [CrossRef]
25. Shinohara, K.; King, C.; Regan, E.J.; Bergman, J.; Carter, A.D.; Arias, A.; Urteaga, M.; Brar, B.; Page, R.; Chaudhuri, R.; et al. GaN-Based Multi-Channel Transistors with Lateral Gate for Linear and Efficient Millimeter-Wave Power Amplifiers. In Proceedings of the 2019 IEEE MTT-S International Microwave Symposium (IMS), Boston, MA, USA, 2–7 June 2019; pp. 1133–1135. [CrossRef]
26. Shinohara, K.; King, C.; Regan, D.; Regan, E.; Carter, A.; Arias, A.; Bergman, J.; Urteaga, M.; Brar, B.; Cao, Y.; et al. Multi-channel Schottky-gate BRIDGE HEMT Technology for Millimeter-Wave Power Amplifier Applications. In Proceedings of the 2022 IEEE/MTT-S International Microwave Symposium-IMS 2022, Denver, CO, USA, 19–24 June 2022; pp. 298–301. [CrossRef]
27. Nagamatsu, K.A.; Afroz, S.; Gupta, S.; Wanis, S.; Hartman, J.; Stewart, E.J.; Shea, P.; Renaldo, K.; Howell, R.S.; Novak, B.; et al. Second Generation SLCFET Amplifier: Improved FT/FMAX and Noise Performance. In Proceedings of the 2019 IEEE BiCMOS and Compound semiconductor Integrated Circuits and Technology Symposium (BCICTS), Nashville, TN, USA, 3–6 November 2019; pp. 1–4. [CrossRef]
28. Azize, M.; Palacios, T. Top-down fabrication of AlGaN/GaN nanoribbons. *Appl. Phys. Lett.* **2011**, *98*, 042103. [CrossRef]
29. Zhang, M.; Ma, X.; Mi, M.; Yang, L.; Wu, S.; Hou, B.; Zhu, Q.; Zhang, H.; Wu, M.; Hao, Y. Influence of fin width and gate structure on the performance of AlGaN/GaN fin-shaped HEMTs. *Int. J. Numer. Model. Electron. Netw. Devices Fields* **2019**, *33*, e2641. [CrossRef]
30. For ATLAS Users' Manual-Device Simulation Software, Silvaco. Available online: http://www.silvaco.com/products/tcad/device_simulation/atlas/atlas.html (accessed on 10 June 2018).
31. Rinke, P.; Winkelnkemper, M.; Qteish, A.; Bimberg, D.; Neugebauer, J.; Scheffler, M. Consistent set of band parameters for the group-III nitrides AlN, GaN, and InN. *Phys. Rev. B* **2008**, *77*, 075202. [CrossRef]
32. Ambacher, O.; Foutz, B.; Smart, J.; Shealy, J.R.; Weimann, N.G.; Chu, K.; Murphy, M.; Sierakowski, A.J.; Schaff, W.J.; Eastman, L.F.; et al. Two dimensional electron gases induced by spontaneous and piezoelectric polarization in undoped and doped AlGaN/GaN heterostructures. *J. Appl. Phys.* **2000**, *87*, 334–344. [CrossRef]
33. Zhang, B.J.; Egawa, T.; Zhao, G.Y.; Ishikawa, H.; Umeno, M.; Jimbo, T. Schottky diodes of Ni/Au on n-GaN grown on sapphire and SiC substrates. *Appl. Phys. Lett.* **2001**, *79*, 2567–2569. [CrossRef]
34. Bisi, D.; Meneghini, M.; de Santi, C.; Chini, A.; Dammann, M.; Bruckner, P.; Mikulla, M.; Meneghesso, G.; Zanoni, E.; Bisi, D.; et al. Deep-Level Characterization in GaN HEMTs-Part I: Advantages and Limitations of Drain Current Transient Measurements. *IEEE Trans. Electron. Devices* **2013**, *60*, 3166–31753. [CrossRef]
35. Ibbetson, J.P.; Fini, P.T.; Ness, K.D.; DenBaars, S.P.; Speck, J.S.; Mishra, U.K. Polarization effects, surface states, and the source of electrons in AlGaN/GaN heterostructure field effect transistors. *Appl. Phys. Lett.* **2000**, *77*, 250–252. [CrossRef]
36. Joshi, V.; Tiwari, S.P.; Shrivastava, M. Part I: Physical Insight into Carbon-Doping-Induced Delayed Avalanche Action in GaN Buffer in AlGaN/GaN HEMTs. *IEEE Trans. Electron. Devices* **2019**, *66*, 561–569. [CrossRef]
37. Maas, S.A. *Nonlinear Microwave and RF Circuits*; Artech House: Norwood, MA, USA, 1997.

Article

Single-Particle Irradiation Effect and Anti-Irradiation Optimization of a JLTFET with Lightly Doped Source

Haiwu Xie [1,2] and Hongxia Liu [1,*]

[1] Key Laboratory for Wide-Band Gap Semiconductor Materials and Devices of Education, The School of Microelectronics, Xidian University, Xi'an 710071, China; xiehaiwu.love@163.com
[2] The School of Physics and Electronic Information Engineering, Qinghai Normal University, Xining 810016, China
* Correspondence: hxliu@mail.xidian.edu.cn; Tel.: +86-189-9723-8105

Abstract: In this article, the particle irradiation effect of a lightly doped Gaussian source heterostructure junctionless tunnel field-effect transistor (DMG-GDS-HJLTFET) is discussed. In the irradiation phenomenon, heavy ion produces a series of electron-hole pairs along the incident track, and then the generated transient current can overturn the logical state of the device when the number of electron-hole pairs is large enough. In the single-particle effect of DMG-GDS-HJLTFET, the carried energy is usually represented by linear energy transfer value (LET). In simulation, the effects of incident ion energy, incident angle, incident completion time, incident position and drain bias voltage on the single-particle effect of DMG-GDS-HJLTFET are investigated. On this basis, we optimize the auxiliary gate dielectric, tunneling gate length for reliability. Simulation results show HfO_2 with a large dielectric constant should be selected as the auxiliary gate dielectric in the anti-irradiation design. Larger tunneling gate leads to larger peak transient drain current and smaller tunneling gate means larger pulse width; from the point of anti-irradiation, the tunneling gate length should be selected at about 10 nm.

Keywords: band-to-band tunneling (BTBT); linear energy transfer value (LET); single-particle irradiation effect; anti-irradiation optimization

Citation: Xie, H.; Liu, H. Single-Particle Irradiation Effect and Anti-Irradiation Optimization of a JLTFET with Lightly Doped Source. *Micromachines* **2023**, *14*, 1413. https://doi.org/10.3390/mi14071413

Academic Editors: Niall Tait and Sadia Ameen

Received: 2 June 2023
Revised: 21 June 2023
Accepted: 12 July 2023
Published: 13 July 2023

1. Introduction

In nanoscale integrated circuits, leakage current increases exponentially with decreasing device feature size and increasing the circuit integration degree. In recent years, among the low-power devices, TFET has attracted a lot of attention due to its own advantages. TFET has small leakage current and good subthreshold characteristics due to the working mechanism of band-to-band tunneling, which is very suitable for low-power circuit applications. Therefore, it is of great significance to study the radiation reliability performance of TFET devices.

The irradiation effect of TFET is less studied in the published literature. At present, Lili Ding [1,2] of Pardova University in Italy has studied the silicon material TFET, and the irradiation source has been chosen as a 10 keV X-ray. The result shows that the oxide trap charge in the gate dielectric changes when the irradiation effect is affected, which leads to the change in threshold voltage and tunneling voltage of the device. The study also compares the Si-based TFET and FDSOI MOSFET, and the result shows that the TFET device has better anti-irradiation characteristics than the FDSOI device. Avashesh Dubey [3,4] conducts a simulation study on Total Ionizing Dose (TID) of SOI TFET, and the results show that the threshold voltage drift and interface trap charge generated by irradiation environment cannot be ignored, and they have a very important effect on the electrical performance of the device. Therefore, the radiation study of TFET has important guiding significance for the practical application of this kind of device [5–10].

With the decreasing feature size of devices, the single-particle irradiation effect gradually becomes the main failure mode of devices in digital applications [11–16]. The reason for this is that the power supply voltage and node-point capacitance decrease with the decreasing feature size in integrated circuits, which leads to the reduction in the storage charge of logic circuits. In other words, the node-point capacitance becomes smaller and smaller, so the circuit is more likely to produce logic errors. At the same time, the decrease in feature size will make the parasitic bipolar effect more serious, and the increase in node spacing and integration will enhance the charge sharing effect after heavy ion impact [17–20].

Generally, the drain bias is set as a fixed value greater than 0 under the OFF-state (Vgs = 0 V) when studying the single-particle transient effect; therefore, the drain terminal voltage is usually larger than other voltages, and there is an inverse PN junction in a P-type TFET at the drain terminal. The strong electric field near the anti-biased junction tends to intensify the single-particle transient effect. When the charged particle is incident, it generates transient current and collected charge at the electrode. If this phenomenon occurs in a storage unit, the collected charge is easy to cause logical upset of the storage unit, that is, single-event upset (SEU) is easy to occur when the collected charge is large enough. Therefore, it is of great practical value to study the transient effects of single particles on the TFET structure [20–24].

In this article, we study the single-particle irradiation effect of the DMG-GDS-HJLTFET structure which has been published in reference [4]. As discussed in reference [4], DMG-GDS-HJLTFET adopts a Gaussian doped source and an InAs/$GaAs_{0.1}Sb_{0.9}$ heterostructure, yet an engineered control gate is designed to improve the ON-state current and suppress the OFF-state current simultaneously by dividing the control gate into two parts, namely the tunnel gate (TG) and the auxiliary gate (AG) with work functions of Φ_{M1} and Φ_{M2}, where $\Phi_{M1} > \Phi_{M2}$. Therefore, the conduction band and the valance band of the proposed device at source/channel interface are very close to each other, resulting in a smaller tunneling distance in DMG-GDS-HJLTFET. Moreover, TG with the work function of Φ_{M1} can lower the minimum value of the conduction band at the channel region, which further promotes the tunneling probability in DMG-GDS-HJLTFET. Compared with DMG-HJLTFET using the Ge/Si heterostructure in reference [4], the ON-state current of DMG-GDS-HJLTFET is three orders of magnitude higher than that of DMG-HJLTFET, increasing to up to 4.1×10^{-4} A/μm. At the same time, the OFF-state current of DMG-GDS-HJLTFET is only 9.92×10^{-19} A/μm, and the subthreshold swing average value of DMG-GDS-HJLTFET is as low as 12.7 mV/Dec. Therefore, DMG-GDS-HJLTFET can be used in future low-power applications.

The cross-section tangent and structure diagram of DMG-GDS-HJLTFET are shown in Figure 1. The proposed structure adopts dual material gate to improve the ON-state and OFF-state current, and a Gaussian doped source is used to simulate the random fluctuation phenomenon of actual impurity doping. The dual material gate can be realized by the method of article [24], and the symmetrical structure can be fabricated by using either molecular beam epitaxy (MBE) or the Metal–organic chemical vapor deposition (MOCVD) method; the method of MBE is especially profoundly researched for building III-V compound semiconductor devices. Based on these analyses, the process flow of the proposed device is as follows: First, the body of the proposed structure can be formed by MBE; Second, HfO_2 can be deposited at the top of the proposed structure. Then, redundant HfO_2 can be etched and SiO_2 needs to be deposited. Afterwards, the gate of the proposed structure can be fabricated. Third, the HfO_2-SiO_2 junction and the divided gate can be formed at the bottom of the proposed structure with the same method. At last, the contacts are defined. Figure 2 shows the tentative fabrication flow for DMG-GDS-HJLTFET.

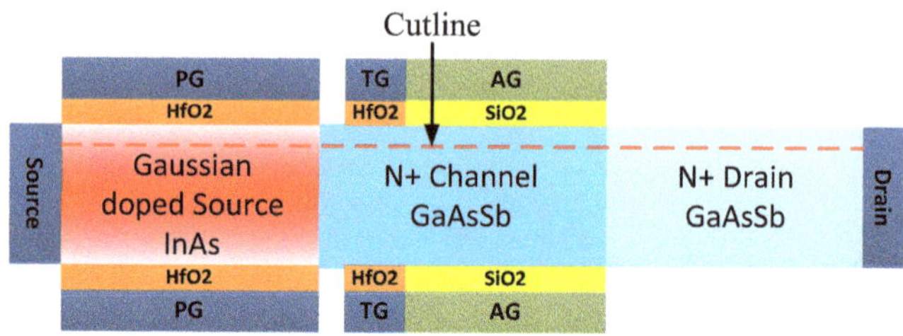

Figure 1. Cross-section tangent and structure diagram of DMG−GDS−HJLTFE.

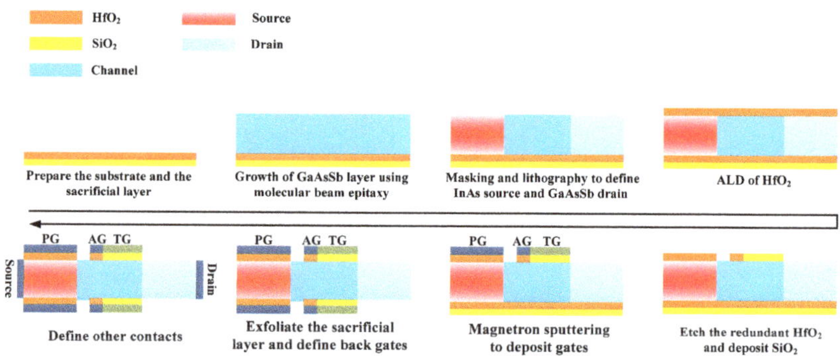

Figure 2. Tentative fabrication flow of DMG−GDS−HJLTFET.

Physical dimensions and electrical properties of DMG-GDS-HJLTFET are discussed in detail in reference [4]. Based on the conclusions in the published literature, we simulate the single-particle irradiation effect of DMG-GDS-HJLTFET in this paper.

All the simulations are carried out using ATLAS Silvaco TCAD version 5.20.2.R. In order to calculate the tunneling current, the nonlocal BTBT model (BBT.NONLOCAL) is activated. A basic analytical formulation for band to band tunneling probability T(E) is shown in Equation (1):

$$T(E) \propto \left(-\frac{4\sqrt{2m^*}{E_g}^{\frac{3}{2}}}{3|e|\hbar\left(E_g+\Delta\Phi\right)} \sqrt{\frac{\varepsilon_{si}}{\varepsilon_{ox}} t_{ox} t_{si}} \right) \Delta\Phi, \quad (1)$$

where m^* is the effective carrier mass, E_g is the bandgap, $\Delta\Phi$ is the energy range over which tunneling can take place, and t_{ox}, t_{Si}, ε_{ox}, and ε_{Si} are the oxide and silicon film thickness and dielectric constants, respectively. As can be seen from the above formulation, small m^* and small E_g are required in source region, and appropriate materials need to be selected in the source/channel interface to ensure the appropriate $\Delta\Phi$.

Shockley–Read–Hall related to concentration (CONSRH) is used to account for the minority carrier recombination effects and the presence of highly doped channel. In this regard, the Fermi Statistics (FERMI) model and the band gap narrowing (BGN) model are included. Quantum confinement model given by Hansch (HANSCHQM) is used to consider the increased doping levels and thinner gate oxide in the channel. The Schenk model for trap-assisted tunneling (SCHENK.TUNN) is used to include the important role of trap-assisted tunneling.

Section 2 introduces single-particle irradiation effect of DMG-GDS-HJLTFET. Section 3 shows the anti-irradiation optimization for DMG-GDS-HJLTFET. Section 4 concludes the paper.

2. Single-Particle Irradiation Effect of DMG-GDS-HJLTFET

The main components of cosmic radiation are about 83% protons, about 13% alpha particles, about 1% heavy ions, and about 3% galactic cosmic radiation electrons and mesons. The energy of these particles needs to be converted when heavy ion incidence occurs on devices of different sizes. In radiation studies, the ionizing particle is typically described by the linear charge deposition (LCD) value; another common measure of the loss of energy is the linear energy transfer (LET) value. The conversion factor from the LET value to the LCD value is then approximately 0.01 for silicon. So, for instance, an LET value of 25 MeV·cm^2/mg is equivalent to 0.25 pC/μm. In this paper, we use a different LET value to represent different types of particles, and the LET value can be set for different particles according to the device structure.

In the irradiation phenomenon, heavy ion incidence produces a series of electron-hole pairs along the track, and the transient current generated with these electron-hole pairs is heavily influenced by linear energy transfer value (LET). Next, we discuss the influence of different LETs. It is worth emphasizing that the OFF-state single-particle effect is more serious than that of the ON-state, so the simulation bias is the device OFF-state, i.e., Vgs = 0 V (gate-to-source voltage) and Vds = 0.5 V (drain-to-source voltage). The default condition of single-particle effect simulation for different LETs is that the incident completion time is equal to 2 ps, the incident angle is equal to 90° and the incident position is the auxiliary gate of the device.

Figure 3a shows the transient drain current of DMG-GDS-HJLTFET with a different LET value when Vgs = 0 V and Vds = 0.5 V and the incident position is TG. The peak value of the transient drain current increases with the increase in LET. The peak value of transient drain current is 1.45×10^{-5} A/μm when LET = 1 MeV·cm^2/mg, and the peak value of transient drain current reaches 7.39×10^{-5} A/μm when LET = 10 MeV·cm^2/mg, which is five times higher than that of 1 MeV·cm^2/mg. At the same time, the pulse width of the transient drain current increases with the LET value. If calculated at 90% of the peak value, the pulse widths of LET = 1 MeV·cm^2/mg and LET = 10 MeV·cm^2/mg are 1.58 ps and 2.77 ps, respectively.

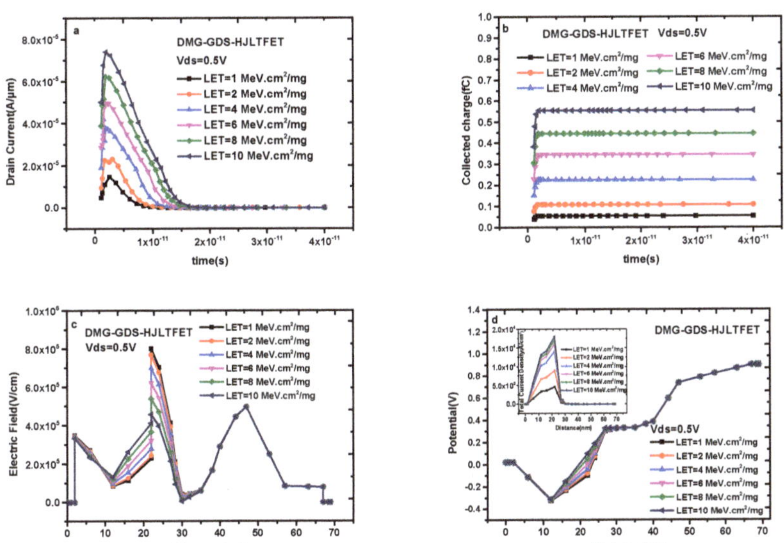

Figure 3. (**a**) Transient drain current of DMG−GDS−HJLTFET; (**b**) transient collected charge of DMG−GDS−HJLTFET; (**c**) change of electric field along the cutline and (**d**) change in potential along the cutline with different LET.

Figure 3b shows the transient collected charge of DMG-GDS-HJLTFET with different LET values where the collected charge is obtained by integrating the transient current during the whole simulation time and the transient current occurs in the form of a pulse. The pulse current drops to 0 at 2×10^{-11} s, so the value of integral is mostly contributed by the current before the 1×10^{-11} s time, and the collected charge stays the same after a specific time. The time for the collected charge to enter the saturation value decreases with the increase in the LET value, and the corresponding time of LET = 10 MeV·cm^2/mg is only 2.78 ps. Moreover, the saturation value of the collected charge increases with the increase in the LET value; the corresponding value of LET = 10 MeV·cm^2/mg is 0.55 fC.

Figure 3c,d, respectively, show the change in electric field and potential along the cutline. The electric field increases with the increase in the LET value in the range of 10–22 nm and decreases with the increase in the LET value in the range of 22–30 nm, which is consistent with the current change in Figure 3a. The inset in Figure 3d reflects the change in charge density with the LET value, which is exactly the same as the current change in Figure 3a.

In the single-particle effect, the incident angle affects the generation of transient current and the collected charge, and a study is conducted at the tunneling gate to examine this effect, where 0° means the incidence parallel to the tunneling gate and 90° represents the incidence perpendicular to the tunneling gate; the remaining angles increase from small to large in the counterclockwise direction, as shown in the inset of Figure 4a. Figure 4a shows the transient drain current of DMG-GDS-HJLTFET with different incident angles. It can be observed from Figure 4a that the peak value of the transient drain current is the largest in the case of parallel incidence. The reason for this is that parallel incidence affects not only the tunneling gate but also the auxiliary gate in the simulation setting. The influence of the tunneling gate and the auxiliary gate on the drain current is analyzed, respectively, in reference [4]; thus, the situation in Figure 4a appears. Moreover, both the pulse width and the peak value of transient drain current decrease when the incident angle increases from 30° to 90°, because incident angle = 30° corresponds to the largest influence area in this process. The pulse width and transient drain current peak affect the distribution of the collected charge, as shown in Figure 4b, and the saturation value change in collected charge in Figure 4b is consistent with transient drain current in Figure 4a.

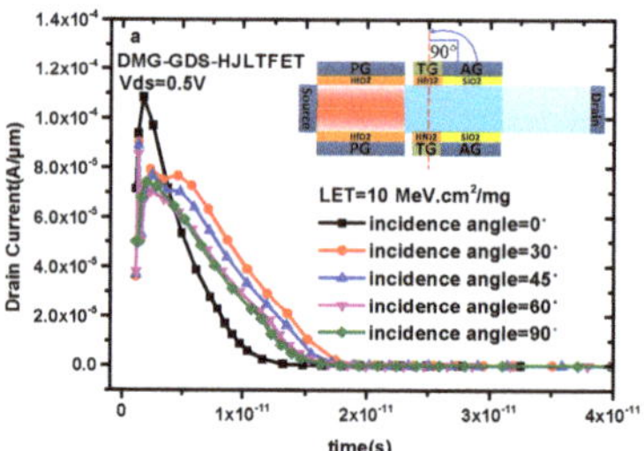

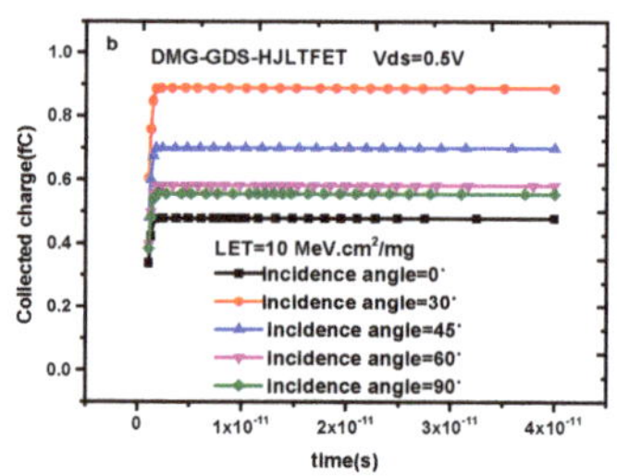

Figure 4. (**a**) Transient drain current of DMG−GDS−HJLTFET and (**b**) transient collected charge of DMG−GDS−HJLTFET with different incident angles (0°, 30°, 45°, 60°, 90°).

Figure 5a,b, respectively, show the variation of transient drain current and collected charge when the incident position changes from the polar gate to the drain region, where the incident position change is depicted in the inset in Figure 5a. On the whole, the closer the incident position is to the drain electrode, the larger the peak value of the corresponding transient current is, which is in accordance with the characteristics of carrier generation and recombination in this device. Unusually, the peak value of the transient drain current is relatively large when the incident location is hetero-dielectric (between AG and TG). The reason for this is that the AG and TG are simultaneously hit at this position, resulting in a large transient value in the current.

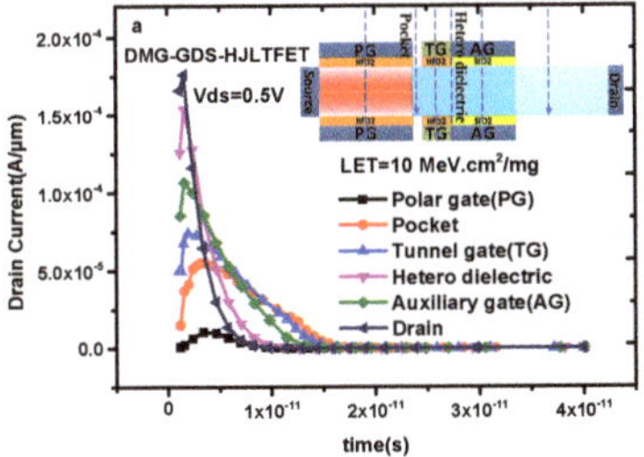

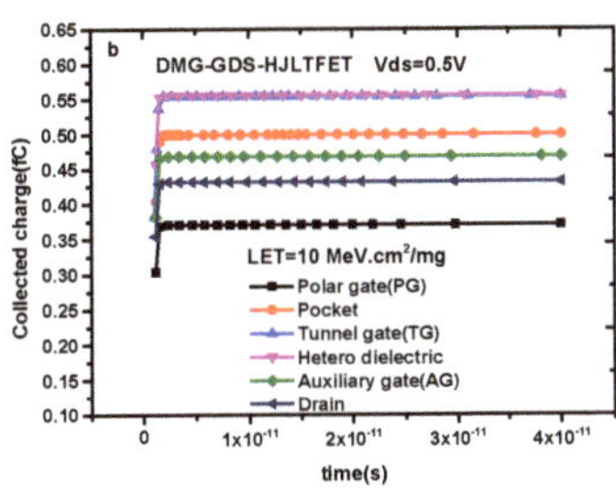

Figure 5. (**a**) Transient drain current of DMG−GDS−HJLTFET and (**b**) transient collected charge of DMG−GDS−HJLTFET with different incident position.

The variation of transient current in Figure 5a determines the change of collected charge in Figure 5b. It can be seen from Figure 5b that the saturation value of the collected charge first increases and then decreases when the incident position is transferred from the polar gate to the drain, and the maximum value of saturation occurs when the incident position is TG.

Figure 6a indicates the transient drain current of DMG-GDS-HJLTFET with different incident completion times where the incident position is TG. Here, the incident completion time goes from 0 ps to 10 ps with the step size of 2 ps. It can be seen that the peak value of the transient drain current increases before tp = 6 ps and obviously attenuates after tp = 8 ps; then, it returns to the level before incidence when tp = 10 ps (tp = 0 ps represents no single-particle irradiation). This change is reflected by a potential variation in Figure 6b. The scope of 12 nm–30 nm is the region of tunneling and tunneling gate in this device, and more non-equilibrium carriers can be generated under greater potential. In addition, long incident completion time means the generated non-equilibrium carriers are more likely to be recombined when they pass through the channel, contributing to the transient drain current changes in Figure 6a.

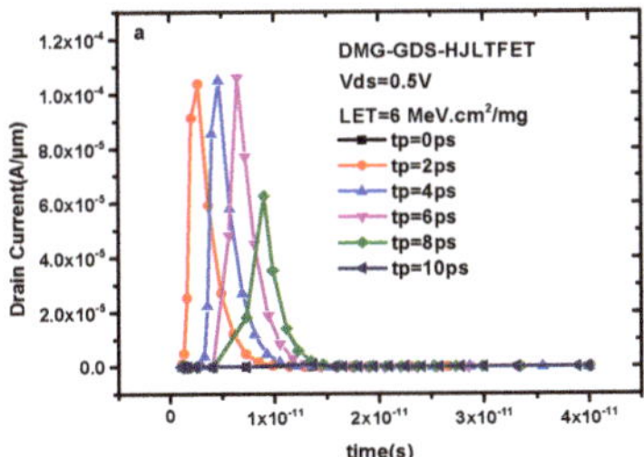

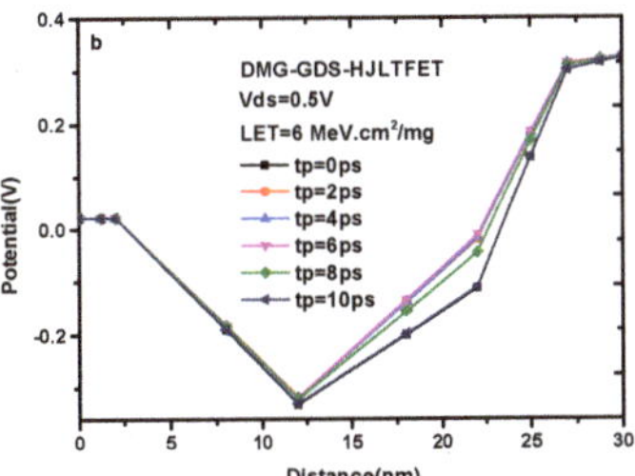

Figure 6. (**a**) Transient drain current of DMG−GDS−HJLTFET and (**b**) corresponding curve of electric potential change with different incident completion time.

Figure 7a shows the influence of drain voltage on transient current, where the LET value is fixed at 10 MeV·cm^2/mg and the incident position is selected as TG. The peak value of transient drain current increases with the increase in Vds, which varies from 5.65×10^{-5} A/μm to 1.23×10^{-4} A/μm when Vds increases from 0.2 V to 0.8 V; however, the pulse width of the transient drain current decreases with the increase in Vds, which decreases from 2.89 ps to 0.4 ps. The reason for this phenomenon is that high drain voltage means high channel potential. More non-equilibrium carriers can be produced in the channel when the drain voltage is higher. At the same time, the generated non-equilibrium carriers are more likely to be collected by the drain voltage at higher values, which will affect the amount of charge collected by the gate. Figure 7b shows the influence of Vds on the collected charge. It can be seen that the non-equilibrium carrier generation process caused by the channel potential is dominant when Vds $\leq$ 0.6 V, while the drain electrode collection process is superior when Vds $\geq$ 0.6 V.

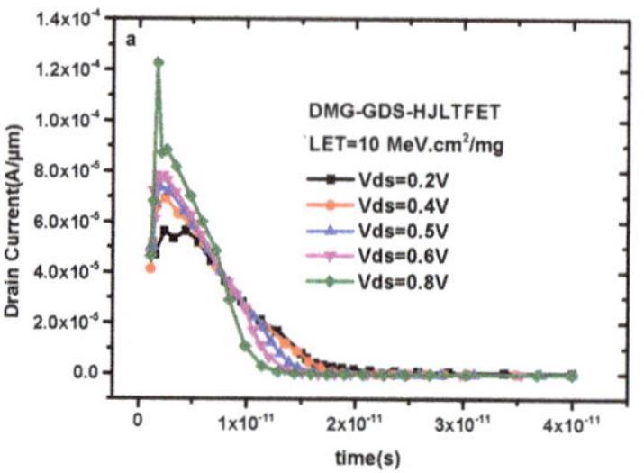

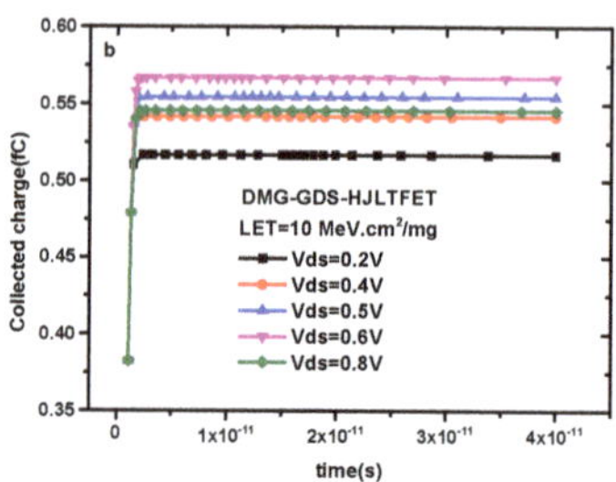

Figure 7. (**a**) Transient drain current of DMG−GDS−HJLTFET and (**b**) transient collected charge of DMG−GDS−HJLTFET with different drain voltage.

3. Anti-Irradiation Optimization for DMG-GDS-HJLTFET

In the previous part, the influence of incident angle, incident position, incident completion time and drain bias voltage on the single-particle irradiation effect of DMG-GDS-HJLTFET is analyzed in detail. The results show that the impact of drain region incidence and auxiliary gate incidence is the most obvious, while the doping concentration in the drain region cannot be changed; therefore, we adjust the type of the auxiliary gate dielectric and the length of the auxiliary gate for DMG-GDS-HJLTFET anti-irradiation optimization.

Figure 8a shows the transient drain current of DMG-GDS-HJLTFET with different auxiliary gate dielectric. It is clear that the dielectric type under the auxiliary gate has an obvious influence on the peak value and pulse width of the transient drain current. The peak value decreases and the pulse width increases with the increase in the dielectric constant. The details are shown in Table 1.

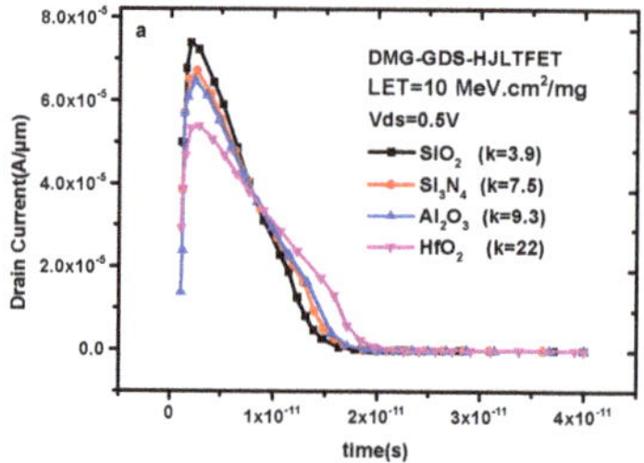

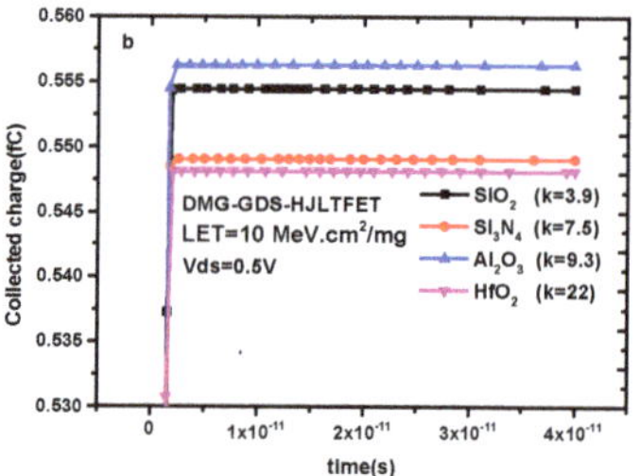

Figure 8. (**a**) Transient drain current of DMG−GDS−HJLTFET and (**b**) transient collected charge of DMG−GDS−HJLTFET with different auxiliary gate dielectric.

Table 1. Comparison of transient drain current and collected charge of DMG-GDS-HJLTFET with different auxiliary gate dielectric.

Dielectric Type	SiO_2	Si_3N_4	Al_2O_3	HfO_2
Dielectric constant	3.9	7.5	9.3	22
Peak value (A/μm)	7.39×10^{-5}	6.73×10^{-5}	6.48×10^{-5}	5.41×10^{-5}
Pulse duration (s)	2.59×10^{-11}	2.64×10^{-11}	2.67×10^{-11}	2.91×10^{-11}
Collected charge (fC)	0.55	0.549	0.556	0.548

It can be observed from Table 1 that a larger auxiliary gate dielectric constant helps to prevent the transient current generated by the single-particle irradiation effect. However, a larger auxiliary gate dielectric constant means a wider pulse width, so it is necessary to compare the saturation value of the collected charge. In Figure 8a, the peak value of the transient drain current decreases with the increase in the dielectric constant while the pulse duration increases with the increase in the dielectric constant. Based on these two aspects, the change in collected charge is shown in Figure 8b. In Figure 8b, there is a maximum collected charge saturation value when the auxiliary gate dielectric is Al_2O_3. The reason for this is that a larger peak current and a wider pulse width occurs when the dielectric layer is chosen as Al_2O_3. However, the pulse is wider when the auxiliary gate dielectric is HfO_2, which can effectively suppress the peak current, resulting in a smaller saturation value of the collected charge. Therefore, in order to effectively prevent the single = -particle irradiation effect, HfO_2 with a larger dielectric constant should be selected as the auxiliary gate dielectric in the anti-irradiation design.

Figure 9a indicates the transient drain current generated by single-particle irradiation in DMG-GDS-HJLTFET when the tunneling gate length increases with a step size of 2 nm within the range of 3 nm to 17 nm (correspondingly, the auxiliary gate length changes from 17 nm to 3 nm). It can be observed from Figure 9a that a longer tunneling gate can produce a greater peak value of the transient drain current. The peak value of the transient drain current is 3.52×10^{-5} A/μm when TG = 3 nm while the peak value of the transient drain current reaches 1.11×10^{-4} A/μm when TG = 17 nm, which increases by more than three times. However, the change in pulse width is just the opposite, and its value gradually decreases with increasing the length of the tunneling gate. Calculated by 90% of the pulse peak value, the pulse width is 2.82 ps when TG = 3 nm and the pulse width is 1.24 ps when TG = 17 nm.

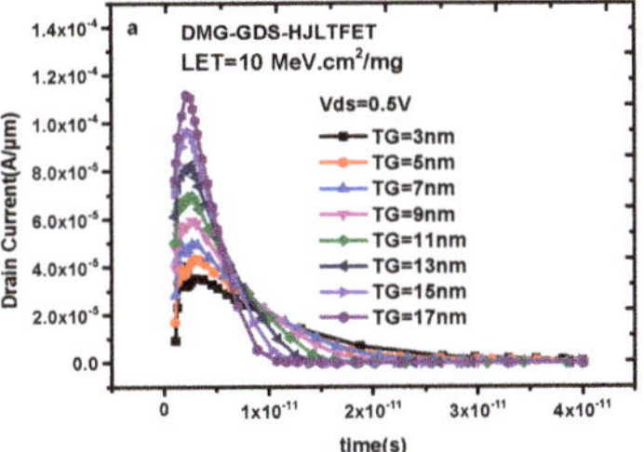

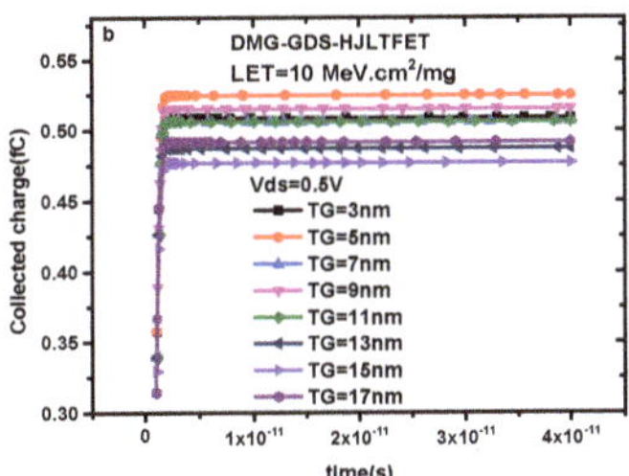

Figure 9. (**a**) Transient drain current of DMG–GDS–HJLTFET and (**b**) transient collected charge of DMG–GDS–HJLTFET with different auxiliary gate length.

Figure 9b shows the collected charge under different tunneling gate lengths. The change in collected charge is related to both the peak value and the pulse width of the transient drain current, that is, the collected charge is the integral of the transient current with respect to time, and its magnitude is positively related to the area formed by the transient drain current and the time axis. The current variation in Figure 9a determines the collected charge distribution in Figure 9b. The saturation value of the collected charge is the largest when TG = 5 nm, which is consistent with the current change in Figure 9a. In the case of single-particle irradiation, a larger tunneling gate leads to a larger peak value of the transient drain current while a smaller tunneling gate means a larger pulse width. Therefore, the appropriate length of the tunneling gate is about 10 nm.

4. Conclusions

In this paper, the single-particle irradiation effect of DMG-GDS-HJLTFET is studied. We discuss the performance under different LETs, different incident angles, different incident completion times, different incident positions and different drain bias voltages in detail. Results show that the peak value of the transient drain current of DMG-GDS-HJLTFET is 7.39×10^{-5} A/μm when LET = 10 MeV·cm^2/mg. If calculated at 90% of the peak value, the pulse widths of LET = 1 MeV·cm^2/mg and LET = 10 MeV·cm^2/mg are 1.58 ps and 2.77 ps, respectively. The pulse width and the peak value of the transient drain current decrease with the increase in the incident angle within 30°–90°, and a closer incident position to the drain electrode can generate a larger peak value of the corresponding transient current. Moreover, the peak value of the transient drain current returns to the level before incidence when tp = 10 ps, and the peak value of the transient drain current increases with the increase in Vds. Based on the comparison of the single-particle irradiation characteristics of DMG-GDS-HJLTFET, the anti-irradiation analysis and optimization are carried out. Results show that HfO_2 with a larger dielectric constant should be selected as the auxiliary gate dielectric in the anti-irradiation design in order to effectively prevent the single-particle irradiation effect, and the appropriate length of the tunneling gate is about 10 nm.

Author Contributions: Conceptualization, H.L. and H.X.; methodology, H.X.; software, H.X.; validation, H.L. and H.X.; formal analysis, H.X.; investigation, H.X.; resources, H.X.; data curation, H.X.; writing—original draft preparation, H.X.; writing—review and editing, H.X.; visualization, H.X.; supervision, H.X.; project administration, H.L.; funding acquisition, H.L. All authors have read and agreed to the published version of the manuscript.

Funding: This research was funded by the National Natural Science Foundation of China (Grant No. U2241221).

Data Availability Statement: The simulation data and the method will be shared upon request to the authors.

Conflicts of Interest: The authors declare no conflict of interest.

References

1. Ding, L.; Gnani, E.; Gerardin, S.; Bagatin, M.; Driussi, F.; Palestri, P.; Selmi, L.; Le Royer, C.; Paccagnella, A. Total Ionizing Dose Effects in Si-Based Tunnel FETs. *IEEE Trans. Electron. Devices* **2014**, *61*, 2874–2880. [CrossRef]
2. Ding, L.; Gnani, E.; Gerardin, S.; Bagatin, M.; Driussi, F.; Selmi, L.; Le Royer, C.; Paccagnella, A. Impact of bias conditions on electrical stress and ionizing radiation effects in Si-based TFETs. *Solid-State Electron.* **2016**, *115*, 146–151. [CrossRef]
3. Dubey, A.; Narang, R.; Saxena, M.; Gupta, M. Investigation of total ionizing dose effect on SOI tunnel FET. *Superlattices Microstruct.* **2019**, *133*, 106186.
4. Xie, H.; Liu, H.; Chen, S.; Han, T.; Wang, S. Electrical performance of $InAs/GaAs_{0.1}Sb_{0.9}$ heterostructure junctionless TFET with dual-material gate and Gaussian-doped source. *Semicond. Sci. Technol.* **2020**, *35*, 95004. [CrossRef]
5. Liu, H.; Cotter, M.; Datta, S.; Narayanan, V. Soft-error performance evaluation on emerging low power devices. *IEEE Trans. Device Mater. Reliab.* **2014**, *14*, 732–741.
6. Hemmat, M.; Kamal, M.; Afzali-Kusha, A.; Pedram, M. Hybrid TFET-MOSFET circuit: A solution to design soft-error resilient ultra-low digital circuit. *Integr. VLSI J.* **2017**, *57*, 11–19. [CrossRef]
7. Sharma, S.; Basu, R.; Kaur, B. Temperature Analysis of a Dopingless TFET Considering Interface Trap Charges for Enhanced Reliability. *IEEE Trans. Electron. Devices* **2022**, *69*, 2692–2697. [CrossRef]
8. Goswami, R.; Bhowmick, B. Comparative Analyses of Circular Gate TFET and Heterojunction TFET for Dielectric-Modulated Label-Free Biosensing. *IEEE Sens. J.* **2019**, *19*, 9600–9608. [CrossRef]
9. Bhattacharyya, A.; De, D.; Chanda, M. Temperature Imposed Sensitivity Issues of Hetero-TFET Based pH Sensor. *IEEE Trans. NanoBioscience* **2022**, *22*, 438–446. [CrossRef]
10. Cheng, Q.; Khandelwal, S.; Zeng, Y. A Physics-Based Model of Vertical TFET—Part II: Drain Current Model. *IEEE Trans. Electron. Devices* **2022**, *69*, 3974–3982. [CrossRef]
11. Song, Y.; Zhang, C.; Dowdy, R.; Chabak, K.; Mohseni, P.K.; Choi, W.; Li, X. III–V Junctionless Gate-All-Around Nanowire MOSFETs for High Linearity Low Power Applications. *IEEE Electron. Device Lett.* **2013**, *35*, 324–326. [CrossRef]
12. Shprits, Y.Y.; Drozdov, A.Y.; Spasojevic, M.; Kellerman, A.C.; Usanova, M.E.; Engebretson, M.J.; Agapitov, O.V.; Zhelavskaya, I.S.; Raita, T.J.; Spence, H.E.; et al. Wave-induced loss of ultra-relativistic electrons in the Van Allen radiation belts. *Nat. Commun.* **2016**, *7*, 12883. [CrossRef]
13. Meyer, P.; Ramaty, R.; Webber, W.R. Cosmic rays-astronomy with energetic particles. *Phys. Today* **1974**, *27*, 23–32. [CrossRef]
14. Badhwar, G.D.; O'Neill, P.M. Galactic cosmic radiation model and its application. *Adv. Sapce Res. Off. J. Comm. Space Res. (COSPAR)* **1996**, *17*, 7–17. [CrossRef]
15. Sweta, C.; Kumar, S.S.; Rekha, C. Comprehensive review on electrical noise analysis of TFET structures. *Superlattices Microstruct.* **2022**, *161*, 107101.
16. Dasika, P.; Watanabe, K.; Taniguch, T.; Majumdar, K. Electrically Self-Aligned, Reconfigurable Test Structure Using $WSe_2/SnSe_2$ Heterojunction for TFET and MOSFET. *IEEE Trans. Electron. Devices* **2022**, *69*, 5377–5381. [CrossRef]
17. Ozturk, M.K. Trajectories of charged particles trapped in Eearth's magnetic field. *Am. J. Phys.* **2011**, *80*, 420–428. [CrossRef]
18. Vanlalawmpuia, K.; Bhowmick, B. Analysis of Hetero-Stacked Source TFET and Heterostructure Vertical TFET as Dielectrically Modulated Label-Free Biosensors. *IEEE Sens. J.* **2022**, *22*, 939–947. [CrossRef]

19. Sathishkumar, M.; Samuel, T.A.; Ramkumar, K.; Anand, I.V.; Rahi, S.B. Performance evaluation of gate engineered InAs–Si heterojunction surrounding gate TFET. *Superlattices Microstruct.* **2022**, *162*, 107099. [CrossRef]
20. Baumann, R.C. Radiation-induced soft errors in advanced semiconductor technologies. *IEEE Trans. Device Mater. Reliab.* **2005**, *5*, 305–316. [CrossRef]
21. Oldham, T.R.; McLean, F.B. Total ionizing dose effects in MOS oxides and devices. *IEEE Trans. Nucl. Sci.* **2003**, *50*, 483–499. [CrossRef]
22. Barnaby, H.J. Total-ionizing-dose effects in modern CMOS technologies. *IEEE Trans. Nucl. Sci.* **2006**, *53*, 3103. [CrossRef]
23. Goyal, P.; Srivastava, G.; Madan, J.; Pandey, R.; Gupta, R.S. Source material valuation of charge plasma based DG-TFET for RFIC applications. *Semicond. Sci. Technol.* **2022**, *37*, 95023. [CrossRef]
24. Long, W.; Ou, H.; Kuo, J.M.; Chin, K.K. Dual-Material Gate (DMG) Field Effect Transistor. *IEEE Trans. Electron. Devices* **1999**, *5*, 865–870. [CrossRef]

MDPI

Article

Argon and Oxygen Gas Flow Rate Dependency of Sputtering-Based Indium-Gallium-Zinc Oxide Thin-Film Transistors

Youngmin Han †, Dong Hyun Lee †, Eou-Sik Cho , Sang Jik Kwon * and Hocheon Yoo *

Department of Electronic Engineering, Gachon University, Seongnam 13120, Republic of Korea
* Correspondence: sjkwon@gachon.ac.kr (S.J.K.); hyoo@gachon.ac.kr (H.Y.)
† These authors contributed equally to this work.

Abstract: Oxygen vacancies are a major factor that controls the electrical characteristics of the amorphous indium-gallium-zinc oxide transistor (a-IGZO TFT). Oxygen vacancies are affected by the composition ratio of the a-IGZO target and the injected oxygen flow rate. In this study, we fabricated three types of a-IGZO TFTs with different oxygen flow rates and then investigated changes in electrical characteristics. Atomic force microscopy (AFM) was performed to analyze the surface morphology of the a-IGZO films according to the oxygen gas rate. Furthermore, X-ray photoelectron spectroscopy (XPS) analysis was performed to confirm changes in oxygen vacancies of a-IGZO films. The optimized a-IGZO TFT has enhanced electrical characteristics such as carrier mobility (μ) of 12.3 $cm^2/V{\cdot}s$, on/off ratio of 1.25×10^{10} A/A, subthreshold swing (*S.S.*) of 3.7 V/dec, and turn-on voltage (V_{to}) of −3 V. As a result, the optimized a-IGZO TFT has improved electrical characteristics with oxygen vacancies having the highest conductivity.

Keywords: a-IGZO; magnetron sputtering; thin-film transistors; oxygen vacancy; oxygen flow rate

Citation: Han, Y.; Lee, D.H.; Cho, E.-S.; Kwon, S.J.; Yoo, H. Argon and Oxygen Gas Flow Rate Dependency of Sputtering-Based Indium-Gallium-Zinc Oxide Thin-Film Transistors. *Micromachines* **2023**, *14*, 1394. https://doi.org/10.3390/mi14071394

Academic Editors: Zeheng Wang and Jingkai Huang

Received: 25 June 2023
Accepted: 6 July 2023
Published: 8 July 2023

1. Introduction

Since 2004, when Hideo Hosono presented an amorphous indium gallium zinc oxide (a-IGZO) and its use in transparent and flexible thin-film TFTs (TFTs), a-IGZO-based TFTs have drawn considerable attention [1–3]. In particular, a-IGZO-based TFTs are currently used in the display industry as this device offers high carrier mobility and, accordingly, sufficient driving current density to operate an organic light-emitting diode (OLED) can be made available [4–6]. Furthermore, another merit of a-IGZO-based TFTs is large-area deposition with high uniformity [7–9]. Due to these merits, many efforts have been made to implement a new concept of electronic devices such as neuromorphic devices [10–12], gas sensors [13–15], photodetectors [16–18], biosensors [19–21], and logic circuits [22–24].

One of the most frequently used deposition methods for a-IGZO is sputtering. This deposition method provides a facile deposition of thin films, particularly oxides, by means of sputtering from a "target" source to a "substrate" [25–27]. As oxygen vacancy concentration in a-IGZO significantly determines the electrical properties of TFTs, specific deposition conditions, including gas flow rate dependency, should be considered [28], and thus, its optimization should be accompanied.

In addition, annealing processes of metal oxide semiconductors (MOS) are improving carrier mobility [29–31]. Oxygen vacancy generated in MOS crystallized through the annealing process are important factors determining conductivity [32,33]. The oxygen vacancy behaves as an electron donor through a fully occupied defect state. Therefore, the conductivity of MOS increases due to the closer Fermi level and conduction band [34,35]. Compared to chemical doping methods and new designs of MOS-based devices, the annealing process provides a simple method and immediate effect for controlling the electrical characteristics of MOS. However, the high annealing temperature of over 600 °C and the additional pre-processing variables reduce the process compatibility of MOS-based devices [36–38].

Here, we investigate the a-IGZO TFT process conditions by controlling the oxygen flow rate without an additional annealing process. The effect of oxygen vacancies controlled by the oxygen gas rate was investigated using atomic force microscopy (AFM) and X-ray photoelectron spectroscopy (XPS) analysis. The a-IGZO TFT fabricated under optimized process conditions exhibits excellent electrical characteristics due to increased conductivity with oxygen vacancies. The carrier mobility and on/off ratio of the optimized a-IGZO TFT are 12.3 $cm^2/V{\cdot}s$ and 1.25 $\times$ 10^{10} A/A, respectively. Also, the subthreshold swing (*S.S.*) of 3.7 V/dec and a V_{to} of -3 V were achieved.

2. Materials and Methods

To fabricate the a-IGZO TFTs, a heavily boron-doped p-type Si/SiO_2 (300 nm) was prepared. The SiO_2 layer was used as the gate dielectric, and a p-type Si layer was applied as the back gate. The Si/SiO_2 wafer was cleaned with acetone and isopropyl alcohol and then dried with nitrogen gas. a-IGZO (In_2O_3: Ga_2O_3: ZnO = 1:1:1) for the channel of TFTs was deposited using the radio frequency (RF) magnetron sputtering method. Condition 1 (C_1) is the deposition method of a-IGZO using only the Ar gas. The gas mixing ratio of O_2:Ar injected during the deposition of a-IGZO is 1.7:100 (Condition 2, C_2) and 17:100 (Condition 3, C_3). The Ti (60 nm) source and drain electrode had deposited electron-beam evaporation. The width (*W*) and length (*L*) of the channels are 100 μm and 1000 μm, respectively. Figure 1a shows the channel and electrode patterning process of a-IGZO TFT using a shadow mask. Figure 1b shows a top-view optical microscope (OM) image of the a-IGZO TFT array. Also, Figure 1c shows the cross-sectional view image of a-IGZO TFT with a scanning electron microscope (SEM), and the thickness of optimized a-IGZO is 8 nm.

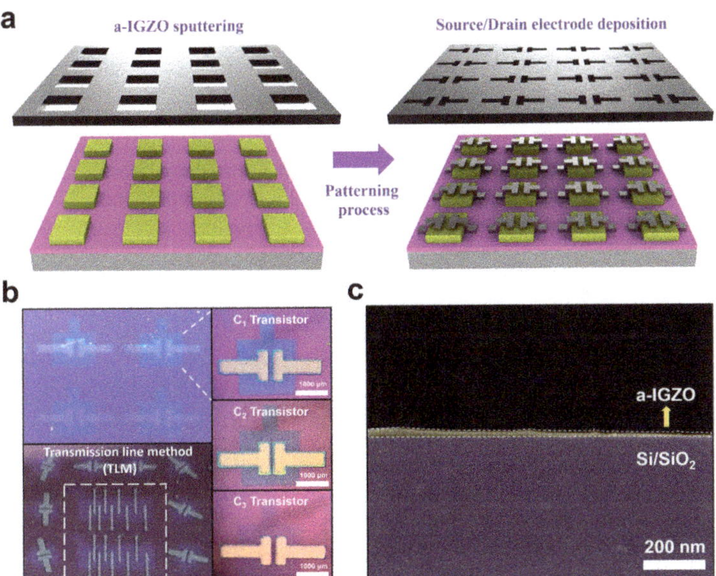

Figure 1. (**a**) 3D schematic of channel and electrode patterning process of a-IGZO TFT using shadow mask. (**b**) Optical microscope image of a-IGZO TFT array according to a-IGZO sputtering conditions. (**c**) Scanning electron microscope image of a-IGZO TFT deposited under optimized process conditions.

The electrical characteristics of a-IGZO TFT were analyzed using a Keithley 4200 (Tektronix, Beaverton, OR, USA) semiconductor parameter analyzer in the air. The electrical characteristics of a-IGZO TFT were investigated according to oxygen gas injection conditions. Atomic force microscopy (AFM) images were measured by XE7 (Park Systems, Suwon, Republic of Korea). X-ray photoelectron spectroscopy (XPS) was measured using AXIS-SUPRA (Kratos, Manchester, UK) at the Korea Basic Science Institute (KBSI). The field-effect mobility and the subthreshold swing (*S.S.*) of TFTs were calculated using the equation:

$$\mu_{electron} = \frac{\partial I_D}{\partial V_G} \frac{L}{W C_{ox} V_D} \tag{1}$$

$$S.S. = \frac{dV_G}{d(\log_{10} I_D)} \tag{2}$$

where L and W represent the length and width of the channel, and C is the capacitance of the gate insulator. The V_{GS} is the applied gate–source voltage, and I_D is the drain current.

Also, the interface trap density (D_{it}) of the TFT was derived using the following equation [39]:

$$D_{it} = \frac{C_{ox}}{q^2}\left(\frac{S.S. \log(e)}{K_B T/q} - 1\right) \tag{3}$$

where q and T represent the elementary electron charge and absolute temperature, respectively. The K_B is the Boltzmann constant, and e is the dielectric constant.

3. Results

Various variables, such as the atomic composition ratio of the a-IGZO target, O_2/Ar mixed gas ratio, and thin film thickness, affect the electrical characteristics of a-IGZO TFTs. In particular, oxygen vacancies are essential due to controlling the electrical characteristics of a-IGZO TFTs. Moreover, oxygen vacancies on the surface of a-IGZO are affected by oxygen gas flow rates. For this reason, we fabricated three types of TFTs with different oxygen gas flow rates to investigate changes in electrical characteristics. Figure 2a shows the transfer curve of the C_1 TFT fabricated under the optimized process conditions. The channel of the C_1 TFT was deposited without oxygen gas injection. The gate–source voltage of the measured transfer curve was −30 V to 60 V, and a drain voltage of 10 V was applied. The on-current and off-current of the C_1 TFT were measured to 6×10^{-4} A and 4.8×10^{-14} A, respectively, and the calculated on/off ratio was measured to as high as 1.25×10^{10} A/A. Figure 2b shows the output curve of the C_1 TFT presenting conventional n-type operation. To measure the output curve, a range of the gate–source voltage is 0 V to 60 V and a drain voltage of 0 V to 50 V were applied, respectively. The contact resistance of a-IGZO TFT was evaluated using the transmission line method (TLM) method. As shown in Figure 2c, the extracted contact resistance is 0.4 Ω·cm using the TLM method. The contact resistance is related to charge injection, which affects electrical characteristics such as mobility and on/off ratio [40–42]. Therefore, the reduced contact resistance promotes charge injection to the channel, and the C_1 TFT had improved mobility and on/off ratio performance. Figure 2d shows the transfer curves with three types of a-IGZO TFTs in different oxygen gas flow rates. The O_2/Ar mixed gas ratios of the C_2 and C_3 TFTs are 1.7:100 and 17:100, respectively. The on-current of the a-IGZO TFT decreases as the oxygen flow rate increases. The electrical characteristics of the optimized a-IGZO TFT depend on the composition ratio of the a-IGZO target. The low conductivity of a-IGZO TFT fabricated with no oxygen component a-IGZO target is overcome by injecting the oxygen gas [43]. On the other hand, the a-IGZO TFT deposited by injecting additional oxygen gas into the a-IGZO target with an oxygen component has reduced oxygen vacancies, resulting in low conductivity. The on-currents at V_{GS} = 60 V of the C_1 and C_2 TFTs were measured to 4.7×10^{-6} A and 6×10^{-4} A, respectively. The C_2 TFT has 127 times less on-current compared to the C_1 TFT. Therefore, the on/off ratio of the C_2 TFT was 1.24×10^6 A/A, which was decreased compared to that of the C_1 TFT. The C_3 TFT had reduced oxygen vacancy compared to C_1 and C_2 TFTs by injecting the highest amount of oxygen gas during a-IGZO sputtering. The low conductivity of the C_3 TFT is attributed to the reduced carrier density due to excessive oxygen vacancy. Therefore, the C_3 TFT has an average current of 67 pA in the gate–source voltage and ranges from −30 V to 60 V and could not be converted to on-state by the positive gate–source bias voltage.

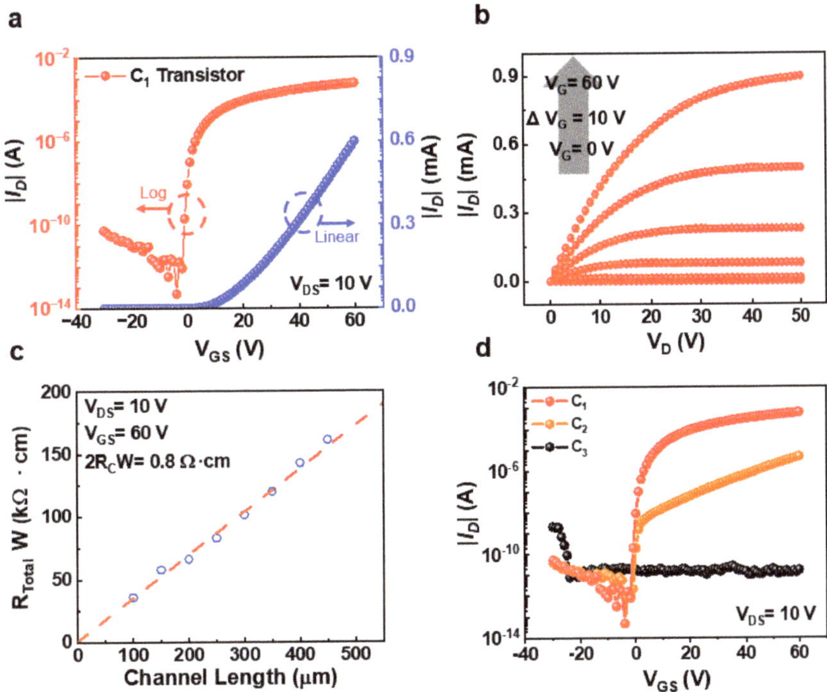

Figure 2. (**a**) Transfer curve of optimized C_1 TFT. (**b**) Output curve of optimized C_1 TFT. (**c**) Contact resistance of optimized C_1 TFT. (**d**) Device electrical characteristics according to oxygen flow variation.

The electrical parameters with the three types of a-IGZO TFTs were compared (Figure 3a–d). The electrical parameters include carrier mobility (μ), turn-on voltage (V_{to}), subthreshold swing (*S.S.*), and on/off ratio. The electrical parameters of the C_3 TFT were uncalculated due to insufficient switching behavior due to the low conductivity of a-IGZO. The carrier mobility of the C_1 TFT and C_2 TFT is 12.3 $cm^2/V{\cdot}s$ and 0.58 $cm^2/V{\cdot}s$, respectively. The carrier mobility of the C_1 TFT is 21 times higher than that of the C_2 TFT. Also, the V_{to} is −3 V and −5 V, respectively, and *S.S.* is 3.7 V/dec and 4.9 V/dec, respectively. As a result, the optimized C_1 TFT achieved a high on/off ratio and carrier mobility. In addition, the electrical characteristics of C_1 transistors with V_{to} close to zero gate–source voltage and low *S.S.* potentially enable low voltage operation and low power consumption. The high electron mobility and reduced *S.S.* of C_1 TFT compared to C_2 TFT are achieved due to the reduced oxygen vacancy and interface trap density (D_{it}) in a-IGZO. The extracted D_{it} of the C_1 TFT and C_2 TFT was 4.4×10^{12} cm^{-2} eV^{-1} and 5.8×10^{12} cm^{-2} eV^{-1}, respectively. Appropriate oxygen vacancy and interface trap density minimize the trapping of charge carriers in C_1 TFT, thereby increasing charge carrier mobility. Also, they decrease the *S.S.*, inducing clear switching behavior of a-IGZO TFT [44–46]. Therefore, optimizing oxygen vacancy and interface trap density is crucial for improving the performance of a-IGZO TFTs, leading to enhanced charge carrier mobility and reduced *S.S.*

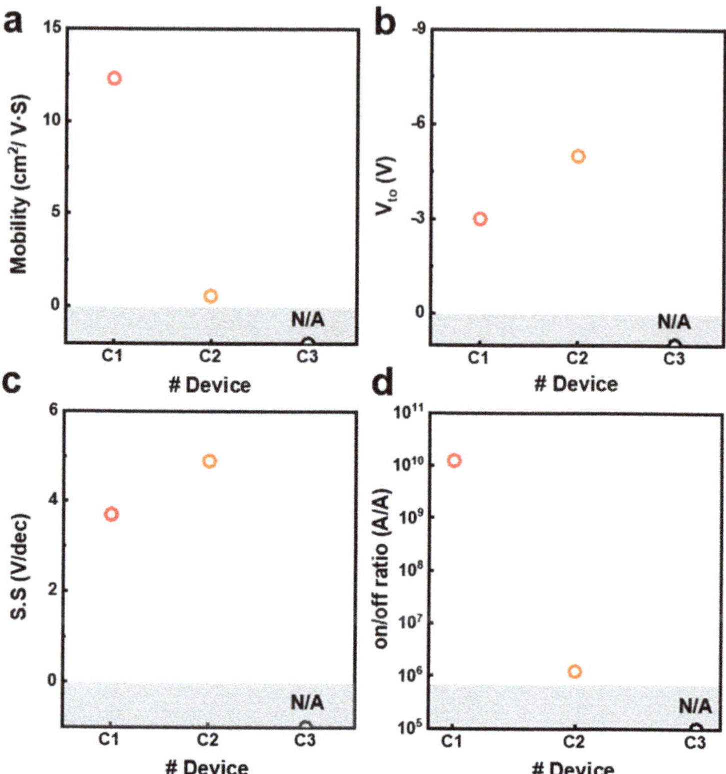

Figure 3. Parameters in three types of a-IGZO TFTs. (**a**) Carrier mobility (μ). (**b**) Turn-on voltage (V_{to}). (**c**) Subthreshold swing (S.S). (**d**) On/off ratio.

Next, we performed an atomic force microscope (AFM) analysis to investigate the morphological characteristics of a-IGZO films fabricated by oxygen flow rate. Figure 4a–c shows the AFM image (5 μM × 5 μM) and height information of a-IGZO films using fabrication methods with C_1, C_2, and C_3, respectively. The height information of the a-IGZO surface did not become rough when oxygen gas flow was increased [47,48]. The morphological characteristics changes on the surface of a-IGZO are related to the concentration of oxygen vacancy. The defect, such as oxygen vacancy, induces rough surface morphology of a-IGZO films. As a result, a high concentration of oxygen vacancy causes the surface of a-IGZO morphology to be rough, while a low concentration of oxygen vacancy induces a smoother surface morphology [31,49]. In addition, the average roughness (R_a) and root mean square roughness (R_q) were calculated from the height information of a-IGZO using the AFM measurement (Figure 4d). The R_a value of C_1, C_2, and C_3 was 0.134, 0.111, and 0.089 nm, respectively. In addition, the values of R_q are 0.151, 0.134, and 0.126 nm, increasing with the oxygen flow rate.

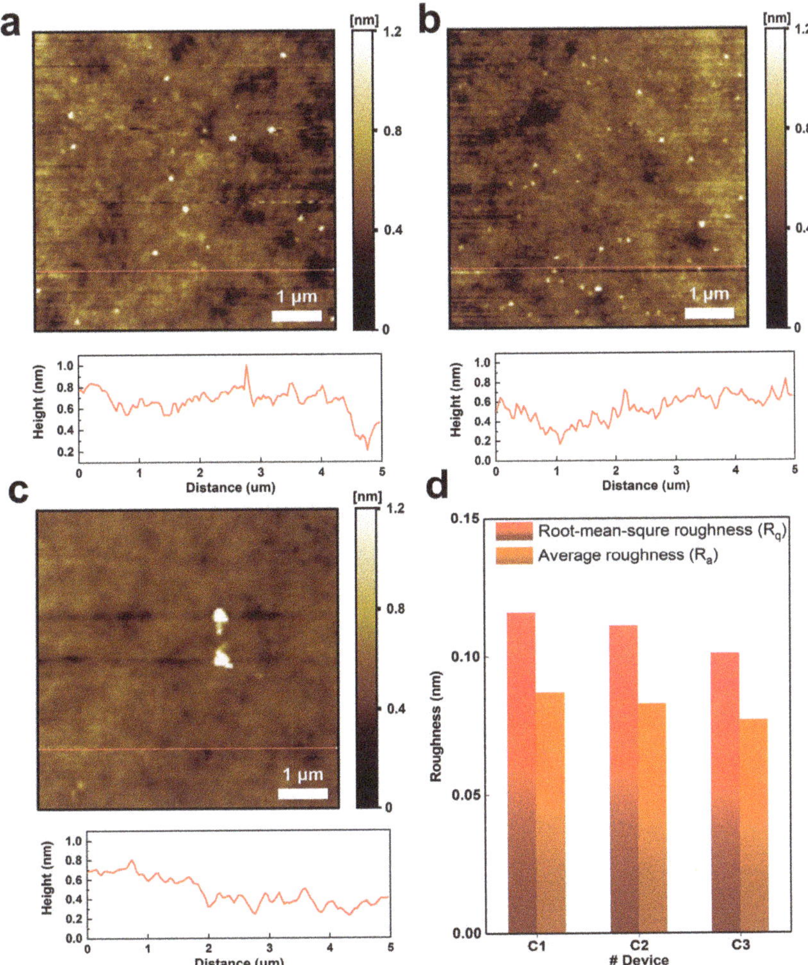

Figure 4. Atomic force microscopy (AFM) image and height information for morphology analysis (**a**) C_1 TFT. (**b**) C_2 TFT. (**c**) C_3 TFT. (**d**) Average roughness (R_a) and root mean square roughness (R_q) of three IGZO films with different oxygen flow rates.

To investigate the chemical composition difference between the three types of a-IGZO films, X-ray photoelectron spectroscopy (XPS) analysis was performed. Figure 5a–c shows the O 1s spectra of the a-IGZO films according to the oxygen gas flow during deposition. The O 1s spectra of the a-IGZO are separated into sub-peaks associated with O_2- ions, such as metal-oxide (M-O) bonds, metal-hydroxyl (M-OH) bonds, and oxygen vacancies. The sub-peaks of O 1s indicate binding at 530 eV, 532 eV, and 531 eV, respectively. The M-O bonds (530 eV) are formed by bonding O_2- ions with elements of a-IGZO metals such as indium, gallium, and zinc. Also, the M-OH bonds (532 eV) were referred between the metals of a-IGZO and the hydroxyl group. In addition, the oxygen vacancies (531 eV) are defects of the a-IGZO channels, which were generated by oxygen atoms and are related to the carrier concentration of a-IGZO. The oxygen gas injected during sputtering reduced the oxygen vacancies of the deposited a-IGZO film. The relative areas of oxygen vacancies in C_1, C_2, and C_3 were 30.6%, 29.6%, and 27.4%, respectively, which oxygen vacancies are reduced when injected oxygen gas was increased. The oxygen vacancy is known to have a significant effect on the change of electrical characteristics in a-IGZO thin firm TFTs [50]. Adequate oxygen vacancies increase the conductivity of a-IGZO by generating free electrons [51]. In addition, the M-OH bond of a-IGZO was decreased as the oxygen flow rate increased [52]. The relative area of M-OH bond in C_1, C_2, and C_3 were 9.5%, 9%, and 4.95%, respectively. The M-OH bond acts as a deep-level trap that prevents charge transport in the a-IGZO, inducing the operation of the unstable TFT [53]. As a result, the optimized C_1 TFT has improved electrical characteristics due to adequate oxygen vacancies and reduced M-OH bonding.

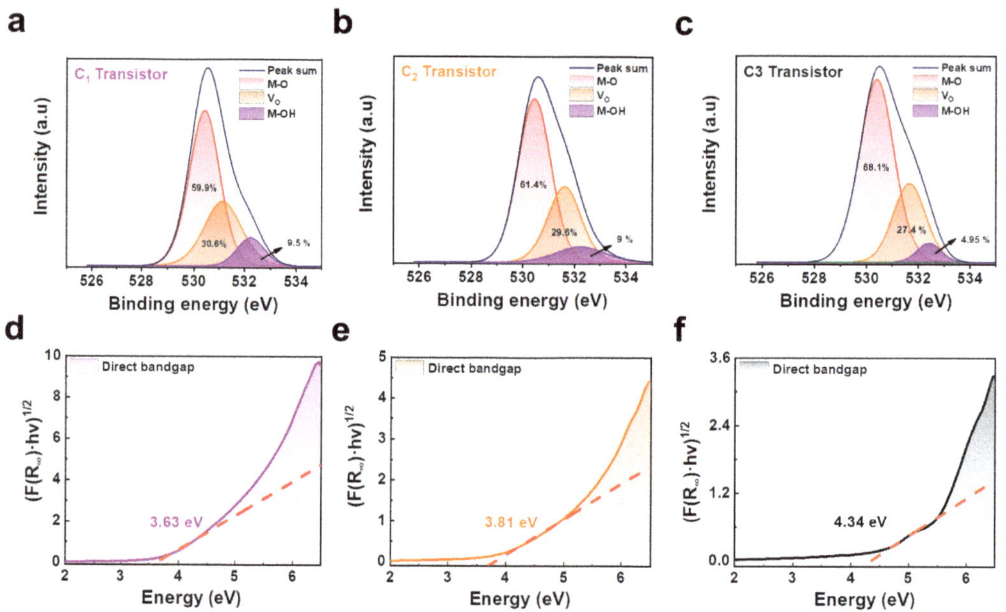

Figure 5. The O 1s peak of XPS spectra in a-IGZO films (**a**) a-IGZO films deposited in C_1. (**b**) a-IGZO films deposited in C_2. (**c**) a-IGZO films deposited in C_3. Optical bandgap of a-IGZO films using Tauc plot method (**d**) a-IGZO films deposited in C_1. (**e**) a-IGZO films deposited in C_2. (**f**) a-IGZO films deposited in C_3.

Moreover, the changes in the optical bandgap of a-IGZO with oxygen gas flow were investigated. Figure 5d–f shows the optical bandgap of a-IGZO calculated by the Tauc plot method using UV–visible absorption data. The a-IGZO in the C_1 condition without oxygen injection showed the smallest optical bandgap of 3.63 eV. On the other hand, the optical bandgaps of C_2 and C_3 increased by oxygen injection are 3.81 eV and 4.34 eV, respectively. As a result, the optical bandgap of a-IGZO was increased as the injected oxygen gas was increased.

4. Conclusions

In summary, the electrical characteristics of oxygen flow rate injection were investigated. As a result, when oxygen flow rate injection was not performed, the best electrical characteristics such as mobility, V_{to}, and on/off ratio were shown. The decreased oxygen vacancy in a-IGZO caused by an increased oxygen gas flow rate results in a decreased conductivity. The C_1 TFT without oxygen gas injected has enhanced charge carrier mobility due to high oxygen vacancy. In addition, AFM and XPS analyses were performed to confirm the mechanism of the change in electrical properties due to oxygen flow injection. In the AFM analysis, the improvement of electrical properties was confirmed through wide contact with the electrode material because the C_1 device had the greatest roughness.

Finally, through XPS analysis, it was confirmed that the electrical characteristics of the C_1 device with the highest oxygen vacancy were the best. Through this result, we believe that the development of oxide semiconductor research and process condition research will be inexhaustible.

Author Contributions: H.Y. contributed to the conceptualization of this research. Y.H. and D.H.L. performed the experiments and analysis and wrote the original manuscript. E.-S.C., S.J.K. and H.Y. reviewed and edited the manuscript. All authors have read and agreed to the published version of the manuscript.

Funding: This work was supported in part by the Korea Institute for Advancement of Technology (KIAT) grant funded by the Korean Government (MOTIE) (No. P0012453, The Competency Development Program for Industry Specialist) and partly supported by the National Research Foundation of Korea (NRF) grant funded by the Korean government (MSIT) (No. NRF-2022R1A2C1003076). This work was supported in part by the Technology Innovation Program (00144300, Interface Technology of 3D Stacked Heterogeneous System for SCM-based Process-in-Memory) funded by the Ministry of Trade, Industry and Energy (MOTIE, Korea).

Conflicts of Interest: The authors declare no conflict of interest.

References

1. Nomura, K.; Ohta, H.; Takagi, A.; Kamiya, T.; Hirano, M.; Hosono, H. Room-Temperature Fabrication of Transparent Flexible Thin-Film Transistors Using Amorphous Oxide Semiconductors. *Nature* **2004**, *432*, 488–492. [CrossRef] [PubMed]
2. Li, X.; Geng, D.; Mativenga, M.; Jang, J. High-Speed Dual-Gate a-IGZO TFT-Based Circuits with Top-Gate Offset Structure. *IEEE Electron. Device Lett.* **2014**, *35*, 461–463. [CrossRef]
3. Rahaman, A.; Chen, Y.; Hasan, M.M.; Jang, J. A High Performance Operational Amplifier Using Coplanar Dual Gate A-IGZO TFTs. *IEEE J. Electron. Devices Soc.* **2019**, *7*, 655–661. [CrossRef]
4. Park, Y.; Cho, D.Y.; Kim, R.; Kim, K.H.; Lee, J.W.; Lee, D.H.; Jeong, S.I.; Ahn, N.R.; Lee, W.; Choi, J.B. Defect Engineering for High Performance and Extremely Reliable A-IGZO Thin-Film Transistor in QD-OLED. *Adv. Electron. Mater.* **2022**, *8*, 2101273. [CrossRef]
5. Sodhani, A.; Kandpal, K. Design of Threshold Voltage Insensitive Pixel Driver Circuitry Using A-IGZO TFT for AMOLED Displays. *Microelectron. J.* **2020**, *101*, 104819. [CrossRef]
6. Kim, D.; Kim, Y.; Lee, S.; Kang, M.S.; Kim, D.H.; Lee, H. High Resolution A-IGZO TFT Pixel Circuit for Compensating Threshold Voltage Shifts and OLED Degradations. *IEEE J. Electron. Devices Soc.* **2017**, *5*, 372–377. [CrossRef]
7. Lin, C.-L.; Chang, W.-Y.; Hung, C.-C. Compensating Pixel Circuit Driving AMOLED Display with A-IGZO TFTs. *IEEE Electron. Device Lett.* **2013**, *34*, 1166–1168. [CrossRef]
8. Kim, J.-S.; Byun, J.-W.; Jang, J.-H.; Kim, Y.-D.; Han, K.-L.; Park, J.-S.; Choi, B.-D. A High-Reliability Carry-Free Gate Driver for Flexible Displays Using a-IGZO TFTs. *IEEE Trans. Electron. Devices* **2018**, *65*, 3269–3276. [CrossRef]
9. Lee, G.J.; Kim, J.; Kim, J.-H.; Jeong, S.M.; Jang, J.E.; Jeong, J. High Performance, Transparent a-IGZO TFTs on a Flexible Thin Glass Substrate. *Semicond. Sci. Technol.* **2014**, *29*, 35003. [CrossRef]
10. Kwon, S.M.; Cho, S.W.; Kim, M.; Heo, J.S.; Kim, Y.; Park, S.K. Environment-adaptable Artificial Visual Perception Behaviors Using a Light-adjustable Optoelectronic Neuromorphic Device Array. *Adv. Mater.* **2019**, *31*, 1906433. [CrossRef]
11. Yang, Y.; He, Y.; Nie, S.; Shi, Y.; Wan, Q. Light Stimulated IGZO-Based Electric-Double-Layer Transistors for Photoelectric Neuromorphic Devices. *IEEE Electron. Device Lett.* **2018**, *39*, 897–900. [CrossRef]
12. Jang, Y.; Park, J.; Kang, J.; Lee, S.-Y. Amorphous InGaZnO (a-IGZO) Synaptic Transistor for Neuromorphic Computing. *ACS Appl. Electron. Mater.* **2022**, *4*, 1427–1448. [CrossRef]
13. Yang, D.J.; Whitfield, G.C.; Cho, N.G.; Cho, P.-S.; Kim, I.-D.; Saltsburg, H.M.; Tuller, H.L. Amorphous InGaZnO4 Films: Gas Sensor Response and Stability. *Sens. Actuators B Chem.* **2012**, *171*, 1166–1171. [CrossRef]
14. Chen, P.-L.; Liu, I.-P.; Chen, W.-C.; Niu, J.-S.; Liu, W.-C. Study of a Platinum Nanoparticle (Pt NP)/Amorphous In-Ga-Zn-O (A-IGZO) Thin-Film-Based Ammonia Gas Sensor. *Sens. Actuators B Chem.* **2020**, *322*, 128592. [CrossRef]
15. Zan, H.-W.; Li, C.-H.; Yeh, C.-C.; Dai, M.-Z.; Meng, H.-F.; Tsai, C.-C. Room-Temperature-Operated Sensitive Hybrid Gas Sensor Based on Amorphous Indium Gallium Zinc Oxide Thin-Film Transistors. *Appl. Phys. Lett.* **2011**, *98*, 253503. [CrossRef]
16. Zhang, Y.Y.; Qian, L.X.; Lai, P.T.; Dai, T.J.; Liu, X.Z. Improved Detectivity of Flexible A-InGaZnO UV Photodetector via Surface Fluorine Plasma Treatment. *IEEE Electron. Device Lett.* **2019**, *40*, 1646–1649. [CrossRef]
17. Yamada, N.; Kondo, Y.; Cao, X.; Nakano, Y. Visible-Blind Wide-Dynamic-Range Fast-Response Self-Powered Ultraviolet Photodetector Based on CuI/In-Ga-Zn-O Heterojunction. *Appl. Mater. Today* **2019**, *15*, 153–162. [CrossRef]
18. Huang, C.-Y.; Peng, T.-Y.; Hsieh, W.-T. Realization of a Self-Powered InGaZnO MSM UV Photodetector Using Localized Surface Fluorine Plasma Treatment. *ACS Appl. Electron. Mater.* **2020**, *2*, 2976–2983. [CrossRef]

19. Lee, I.-K.; Lee, K.H.; Lee, S.; Cho, W.-J. Microwave Annealing Effect for Highly Reliable Biosensor: Dual-Gate Ion-Sensitive Field-Effect Transistor Using Amorphous InGaZnO Thin-Film Transistor. *ACS Appl. Mater. Interfaces* **2014**, *6*, 22680–22686. [CrossRef] [PubMed]
20. Son, H.W.; Park, J.H.; Chae, M.-S.; Kim, B.-H.; Kim, T.G. Bilayer Indium Gallium Zinc Oxide Electrolyte-Gated Field-Effect Transistor for Biosensor Platform with High Reliability. *Sens. Actuators B Chem.* **2020**, *312*, 127955. [CrossRef]
21. Bhatt, D.; Kumar, S.; Panda, S. Amorphous IGZO Field Effect Transistor Based Flexible Chemical and Biosensors for Label Free Detection. *Flex. Print. Electron.* **2020**, *5*, 14010. [CrossRef]
22. Seok, M.J.; Choi, M.H.; Mativenga, M.; Geng, D.; Kim, D.Y.; Jang, J. A Full-Swing a-IGZO TFT-Based Inverter with a Top-Gate-Bias-Induced Depletion Load. *IEEE Electron. Device Lett.* **2011**, *32*, 1089–1091. [CrossRef]
23. Cho, I.-T.; Lee, J.-W.; Park, J.-M.; Cheong, W.-S.; Hwang, C.-S.; Kwak, J.-S.; Cho, I.-H.; Kwon, H.-I.; Shin, H.; Park, B.-G. Full-Swing a-IGZO Inverter with a Depletion Load Using Negative Bias Instability under Light Illumination. *IEEE Electron. Device Lett.* **2012**, *33*, 1726–1728. [CrossRef]
24. Kim, J.-S.; Jang, J.-H.; Kim, Y.-D.; Byun, J.-W.; Han, K.; Park, J.-S.; Choi, B.-D. Dynamic Logic Circuits Using A-IGZO TFTs. *IEEE Trans. Electron. Devices* **2017**, *64*, 4123–4130. [CrossRef]
25. Cho, M.H.; Seol, H.; Song, A.; Choi, S.; Song, Y.; Yun, P.S.; Chung, K.-B.; Bae, J.U.; Park, K.-S.; Jeong, J.K. Comparative Study on Performance of IGZO Transistors with Sputtered and Atomic Layer Deposited Channel Layer. *IEEE Trans. Electron. Devices* **2019**, *66*, 1783–1788. [CrossRef]
26. Yao, R.; Zheng, Z.; Xiong, M.; Zhang, X.; Li, X.; Ning, H.; Fang, Z.; Xie, W.; Lu, X.; Peng, J. Low-Temperature Fabrication of Sputtered High-k HfO2 Gate Dielectric for Flexible a-IGZO Thin Film Transistors. *Appl. Phys. Lett.* **2018**, *112*, 103503. [CrossRef]
27. Yao, R.; Zheng, Z.; Fang, Z.; Zhang, H.; Zhang, X.; Ning, H.; Wang, L.; Peng, J.; Xie, W.; Lu, X. High-Performance Flexible Oxide TFTs: Optimization of a-IGZO Film by Modulating the Voltage Waveform of Pulse DC Magnetron Sputtering without Post Treatment. *J. Mater. Chem. C Mater.* **2018**, *6*, 2522–2532. [CrossRef]
28. Choi, H.-W.; Song, K.-W.; Kim, S.-H.; Nguyen, K.T.; Eadi, S.B.; Kwon, H.-M.; Lee, H.-D. Zinc Oxide and Indium-Gallium-Zinc-Oxide Bi-Layer Synaptic Device with Highly Linear Long-Term Potentiation and Depression Characteristics. *Sci. Rep.* **2022**, *12*, 1259. [CrossRef]
29. Hwang, S.; Lee, J.H.; Woo, C.H.; Lee, J.Y.; Cho, H.K. Effect of Annealing Temperature on the Electrical Performances of Solution-Processed InGaZnO Thin Film Transistors. *Thin Solid. Film.* **2011**, *519*, 5146–5149. [CrossRef]
30. Lin, C.-I.; Yen, T.-W.; Lin, H.-C.; Huang, T.-Y.; Lee, Y.-S. Effect of Annealing Ambient on the Characteristics of A-IGZO Thin Film Transistors. In Proceedings of the 4th IEEE International NanoElectronics Conference, Taoyuan, Taiwan, 21–24 June 2011; pp. 1–2.
31. Peng, C.; Yang, S.; Pan, C.; Li, X.; Zhang, J. Effect of Two-Step Annealing on High Stability of a-IGZO Thin-Film Transistor. *IEEE Trans. Electron. Devices* **2020**, *67*, 4262–4268. [CrossRef]
32. Jeong, S.; Ha, Y.; Moon, J.; Facchetti, A.; Marks, T.J. Role of Gallium Doping in Dramatically Lowering Amorphous-oxide Processing Temperatures for Solution-derived Indium Zinc Oxide Thin-film Transistors. *Adv. Mater.* **2010**, *22*, 1346–1350. [CrossRef]
33. Greiner, M.T.; Chai, L.; Helander, M.G.; Tang, W.; Lu, Z. Transition Metal Oxide Work Functions: The Influence of Cation Oxidation State and Oxygen Vacancies. *Adv. Funct. Mater.* **2012**, *22*, 4557–4568. [CrossRef]
34. Chen, C.; Cheng, K.-C.; Chagarov, E.; Kanicki, J. Crystalline In–Ga–Zn–O Density of States and Energy Band Structure Calculation Using Density Function Theory. *Jpn. J. Appl. Phys.* **2011**, *50*, 091102. [CrossRef]
35. Allen, M.W.; Durbin, S.M. Influence of Oxygen Vacancies on Schottky Contacts to ZnO. *Appl. Phys. Lett.* **2008**, *92*, 122110. [CrossRef]
36. Kim, Y.-M.; Kang, H.-B.; Kim, G.-H.; Hwang, C.-S.; Yoon, S.-M. Improvement in Device Performance of Vertical Thin-Film Transistors Using Atomic Layer Deposited IGZO Channel and Polyimide Spacer. *IEEE Electron. Device Lett.* **2017**, *38*, 1387–1389. [CrossRef]
37. Pu, H.; Zhou, Q.; Yue, L.; Zhang, Q. Investigation of Oxygen Plasma Treatment on the Device Performance of Solution-Processed a-IGZO Thin Film Transistors. *Appl. Surf. Sci.* **2013**, *283*, 722–726. [CrossRef]
38. Yamazaki, S.; Atsumi, T.; Dairiki, K.; Okazaki, K.; Kimizuka, N. In-Ga-Zn-Oxide Semiconductor and Its Transistor Characteristics. *ECS J. Solid. State Sci. Technol.* **2014**, *3*, Q3012. [CrossRef]
39. Abliz, A. Hydrogenation of MG-Doped InGaZno Thin-Film Transistors for Enhanced Electrical Performance and Stability. *IEEE Trans. Electron. Devices* **2021**, *68*, 3379–3383. [CrossRef]
40. Shen, P.-C.; Su, C.; Lin, Y.; Chou, A.-S.; Cheng, C.-C.; Park, J.-H.; Chiu, M.-H.; Lu, A.-Y.; Tang, H.-L.; Tavakoli, M.M. Ultralow Contact Resistance between Semimetal and Monolayer Semiconductors. *Nature* **2021**, *593*, 211–217. [CrossRef]
41. Si, M.; Andler, J.; Lyu, X.; Niu, C.; Datta, S.; Agrawal, R.; Ye, P.D. Indium–Tin-Oxide Transistors with One Nanometer Thick Channel and Ferroelectric Gating. *ACS Nano* **2020**, *14*, 11542–11547. [CrossRef]
42. Wang, W.; Li, L.; Lu, C.; Liu, Y.; Lv, H.; Xu, G.; Ji, Z.; Liu, M. Analysis of the Contact Resistance in Amorphous InGaZnO Thin Film Transistors. *Appl. Phys. Lett.* **2015**, *107*, 063504. [CrossRef]
43. On, N.; Kim, B.K.; Lee, S.; Kim, E.H.; Lim, J.H.; Jeong, J.K. Hot Carrier Effect in Self-Aligned In–Ga–Zn–O Thin-Film Transistors with Short Channel Length. *IEEE Trans. Electron. Devices* **2020**, *67*, 5544–5551. [CrossRef]
44. Abliz, A.; Wan, D.; Chen, J.-Y.; Xu, L.; He, J.; Yang, Y.; Duan, H.; Liu, C.; Jiang, C.; Chen, H. Enhanced Reliability of In–Ga–ZnO Thin-Film Transistors through Design of Dual Passivation Layers. *IEEE Trans. Electron. Devices* **2018**, *65*, 2844–2849. [CrossRef]

45. Abliz, A.; Nurmamat, P.; Wan, D. Rational Design of Oxide Heterostructure InGaZnO/TiO2 for High-Performance Thin-Film Transistors. *Appl. Surf. Sci.* **2023**, *609*, 155257. [CrossRef]
46. Zhou, X.; Han, D.; Dong, J.; Li, H.; Yi, Z.; Zhang, X.; Wang, Y. The Effects of Post Annealing Process on the Electrical Performance and Stability of Al-Zn-O Thin-Film Transistors. *IEEE Electron. Device Lett.* **2020**, *41*, 569–572. [CrossRef]
47. Chen, X.F.; He, G.; Liu, M.; Zhang, J.W.; Deng, B.; Wang, P.H.; Zhang, M.; Lv, J.G.; Sun, Z.Q. Modulation of Optical and Electrical Properties of Sputtering-Derived Amorphous InGaZnO Thin Films by Oxygen Partial Pressure. *J. Alloys Compd.* **2014**, *615*, 636–642. [CrossRef]
48. Ling, L.; Tao, X.; Zhongxiao, S.; Chunliang, L.; Fei, M. Effect of Sputtering Pressure on Surface Roughness, Oxygen Vacancy and Electrical Properties of a-IGZO Thin Films. *Rare Met. Mater. Eng.* **2016**, *45*, 1992–1996. [CrossRef]
49. Zhang, Y.; Zhang, H.; Yang, J.; Ding, X.; Zhang, J. Solution-Processed Yttrium-Doped IZTO Semiconductors for High-Stability Thin Film Transistor Applications. *IEEE Trans. Electron. Devices* **2019**, *66*, 5170–5176. [CrossRef]
50. Tian, Y.; Han, D.; Cai, J.; Geng, Y.; Wang, W.; Wang, L.; Zhang, S.; Wang, Y. Low-Temperature Fabrication of Fully Transparent IGZO Thin Film Transistors on Glass Substrate. In Proceedings of the 2012 IEEE International Conference on Electron Devices and Solid State Circuit (EDSSC), Bangkok, Thailand, 3–5 December 2012; pp. 1–3.
51. Nomura, K.; Takagi, A.; Kamiya, T.; Ohta, H.; Hirano, M.; Hosono, H. Amorphous Oxide Semiconductors for High-Performance Flexible Thin-Film Transistors. *Jpn. J. Appl. Phys.* **2006**, *45*, 4303. [CrossRef]
52. Lee, E.; Kim, T.H.; Lee, S.W.; Kim, J.H.; Kim, J.; Jeong, T.G.; Ahn, J.-H.; Cho, B. Improved Electrical Performance of a Sol–Gel IGZO Transistor with High-k Al_2O_3 Gate Dielectric Achieved by Post Annealing. *Nano Converg.* **2019**, *6*, 1–8. [CrossRef]
53. Zhang, C.; Li, D.; Lai, P.T.; Huang, X.D. An InGaZnO Charge-Trapping Nonvolatile Memory with the Same Structure of a Thin-Film Transistor. *IEEE Electron. Device Lett.* **2021**, *43*, 32–35. [CrossRef]

micromachines

Review

A Review on Micro-LED Display Integrating Metasurface Structures

Zhaoyong Liu [1,2,3], Kailin Ren [1,2,3], Gaoyu Dai [1,2,3] and Jianhua Zhang [1,2,3,*]

1 School of Microelectronics, Shanghai University, Shanghai 200444, China; lzyly@shu.edu.cn (Z.L.); renkailin@shu.edu.cn (K.R.); gaoyudai@shu.edu.cn (G.D.)
2 Key Laboratory of Advanced Display and System Applications (Ministry of Education), Shanghai University, Shanghai 200444, China
3 Shanghai Key Laboratory of Chips and Systems for Intelligent Connected Vehicle, Shanghai University, Shanghai 200444, China
* Correspondence: jhzhang@shu.edu.cn

Abstract: Micro-LED display technology has been considered a promising candidate for near-eye display applications owing to its superior performance, such as having high brightness, high resolution, and high contrast. However, the realization of polarized and high-efficiency light extraction from Micro-LED arrays is still a significant problem to be addressed. Recently, by exploiting the capability of metasurfaces in wavefront modulation, researchers have achieved many excellent results by integrating metasurface structures with Micro-LEDs, including improving the light extraction efficiency, controlling the emission angle to achieve directional emission, and obtaining polarized Micro-LEDs. In this paper, recent progressions on Micro-LEDs integrated with metasurface structures are reviewed in the above three aspects, and the similar applications of metasurface structures in organic LEDs, quantum dot LEDs, and perovskite LEDs are also summarized.

Keywords: Micro-LED; metasurface; light extraction efficiency; angular deflection; polarization

Citation: Liu, Z.; Ren, K.; Dai, G.; Zhang, J. A Review on Micro-LED Display Integrating Metasurface Structures. *Micromachines* **2023**, *14*, 1354. https://doi.org/10.3390/mi14071354

Academic Editors: Zeheng Wang and Jingkai Huang

Received: 9 June 2023
Revised: 27 June 2023
Accepted: 29 June 2023
Published: 30 June 2023

1. Introduction

Micro-LEDs have attracted much attention owing to their advantages of high luminous intensity, high resolution, high contrast, fast response speed, long lifespan, and low power consumption. Due to these excellent performance traits, Micro-LEDs are regarded as the mainstream of next-generation display technology for a wide range of applications, from wearable devices such as wristbands and watches to commercial billboards, public displays, and virtual reality (VR) or augmented reality (AR) devices [1–5]. However, challenges have also arisen with the development of Micro-LED display technology, such as mass transfer, full-color display, and size-dependent efficiency [6,7]. The luminous efficiency of Micro-LEDs decreases rapidly as the size decreases, so it is necessary to improve the light extraction efficiency (LEE) to improve the external quantum efficiency (EQE) [8]. Nowadays, there are many methods to improve LEE. This paper mainly reviews the approaches to integrating metasurface structures on Micro-LEDs.

The metasurface is an artificial nanostructure that is designed to control the amplitude, polarization, and phase of incident waves at the subwavelength scale [9–12]. Metasurface structures can realize the above functions with the premise that the incident light must be coherent [13]. However, a typical Micro-LED exhibits Lambertian-shaped emission [14]. Light emitted in any direction has very low spatial coherence, so it is a key issue to realize control of the Micro-LED wavefront with metasurface. Therefore, the researchers introduced reflective mirrors at the bottom and top of the Micro-LED to form a Fabry–Perot (F-P) cavity structure [13], so that the emitted light is concentrated in a narrow angular range after resonance selection through the cavity, which can enhance the spatial coherence of the emitted light, and the collimation of the emitted light also improves the LEE. In this case, the integration with the metasurface structures can realize the deflection of the beam angle, and the light can be emitted to the preset position to fully utilize the emitted light.

In addition, Micro-LEDs that emit polarized light play a key role in near-eye displays, but obtaining polarized emission from LEDs requires complex design and manufacturing [15]. Moreover, obtaining polarized light emission is difficult due to the weak anisotropy of Micro-LEDs, so wave plates are needed to improve the anisotropy of Micro-LEDs. However, traditional wave plates are not conducive to Micro-LED integration due to their large size. The appearance of the metasurface structures solves these problems because of their small size and simple implementation process [16]. Moreover, the combination of metasurface structures and optical gratings can realize linear and circular polarization. This paper reviews the research progress of Micro-LEDs integrated with metasurface structures in improving the LEE, collimation, controlling angle deflection, and controlling polarization. The latest research results in the four directions mentioned above are demonstrated in Figure 1.

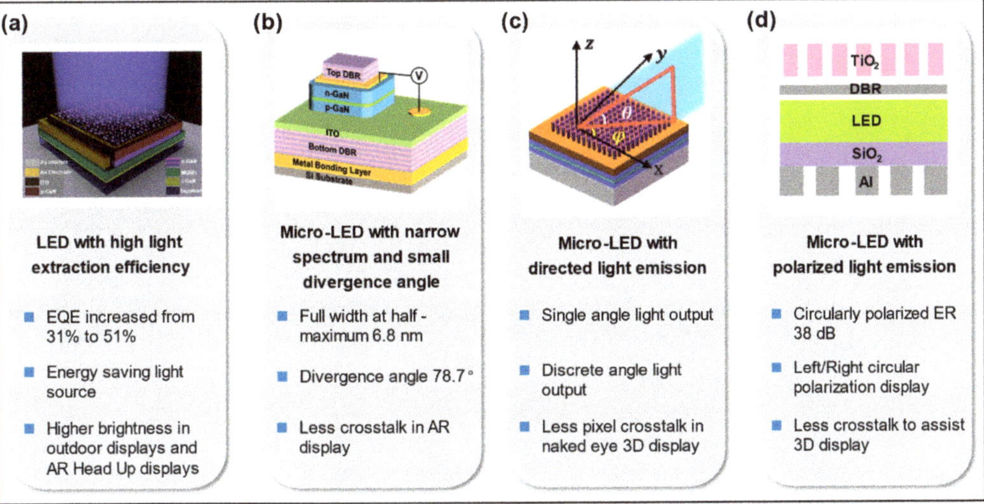

Figure 1. A list of four research directions for LEDs integrated with metasurface. (**a**) Disordered Ag nanoparticle metasurface to improve the LEE of LEDs. Reproduced with permission from [17], [*Light: Science & Applications*]; published by Nature Publishing Group, 2021. (**b**) Distributed Bragg reflector (DBR) resonator to improve the emitted light collimation of Micro-LEDs. Reproduced with permission from [18], [*Applied Physics Letters*]; published by AIP Publishing, 2022. (**c**) Metasurface structures to achieve directed light emission of Micro-LEDs. Adapted with permission from [19] © The Optical Society. (**d**) Metasurface and grating to improve the extinction ratio (ER) of polarized Micro-LEDs. Adapted with permission from [15] © The Optical Society.

2. Improvement in Light Extraction Efficiency

As one of the important performance indicators of Micro-LEDs, EQE refers to the ratio of the final emitted photon number to the injected carrier number, which can be obtained by the product of the LEE and internal quantum efficiency (IQE). Therefore, EQE can be improved by improving the LEE. Currently, methods to improve the LEE of Micro-LEDs include flip chip technology, transparent substrate technology, patterned substrate technology, surface microstructure technology, and bottom reflector technology [20–22]. This paper mainly introduces metasurface structures to improve the LEE and summarizes them in this chapter (listed in Table 1).

Table 1. Summary of research on metasurface structure to improve the LEE of LEDs.

Research Objective	LED Type	Wavelength	Optimized Structure Model	Simulation or Experiment	Ref.
Improve LEE	LED	440~470 nm	Disordered Ag nanoparticles	Simulation and experiment	[17]
Improve LEE	PeLED	780 nm	Nanobricks in the electron transport layer	Simulation	[23]
Improve LEE	OLED	440~570 nm	Speckle image holography metasurfaces	Simulation and experiment	[24]
Improve LEE	OLED	470~630 nm	Reflected supergrating	Simulation and experiment	[25,26]
Improve LEE	OLED	450~650 nm	Bottom nanoslot metasurface	Simulation and experiment	[27]
Improve LEE	Micro-OLED	640 nm	Metalens embedding the glass substrate	Simulation	[28]
Study the influence of wavelength and material on the SIH metasurface to improve LEE	OLED	440~670 nm	SIH metasurfaces	Simulation and experiment	[29]

2.1. Improvement in the LEE by Metasurface Structures

Since the refractive index of the surface material of Micro-LEDs is much larger than that of air, the full reflection angle of the light emitted by Micro-LEDs into the air is very small, causing total reflection phenomena to very easily appear, which is also one of the reasons for the low LEE [30]. The integrated metasurface structures on the top surface of Micro-LEDs can not only effectively expand the light emitting area of Micro-LEDs, but also change the incident angle of light on the inner surface of the semiconductor, thus destroying the total reflection condition of part of the light so as to significantly improve the LEE.

Mao et al. proposed disordered metasurface LEDs by studying the distribution and size of the cuticle fringes in fireflies [17]. The metasurface structure changed from an orderly arrangement of square gratings to a curved top surface, then to a disordered arrangement, and was finally designed as Ag nanoparticles with a curved top surface and disordered arrangement (see Figure 2a). LEDs with Ag-free nanoparticles were compared with those with Ag nanoparticles. Figure 2b,c show the photoluminescence (PL) and electroluminescence (EL) spectra with and without Ag nanoparticles, respectively. The results showed that Ag nanoparticles enhanced PL and EL by 170% and 140%, respectively. The wavelength of the experiment in this paper can also be obtained from Figure 2b,c by taking the full width at half-maximum (FWHM); the wavelength is found to be 440~470 nm, and it is listed in Table 1. The wavelengths of all tables in this review are derived from this. When the working wavelength is 452 nm, the output power of the LED with Ag nanoparticles is increased by 2.7 times, and the absolute EQE is increased from 31% to 51% (see Figure 2d,e). In addition, the top metasurface structure can be made in one step by the gas cluster technology, which reduces the complexity of the process and has a good application prospect.

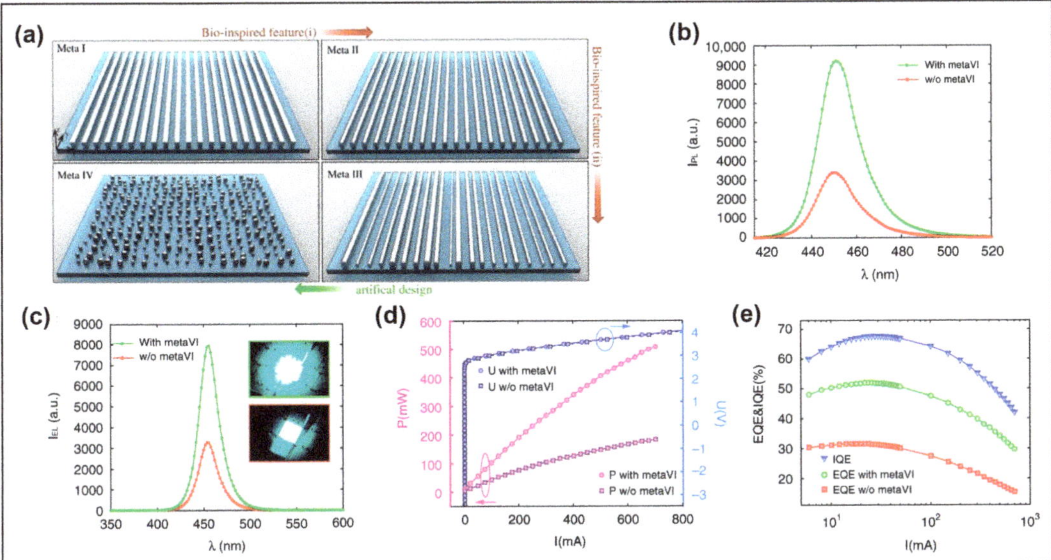

Figure 2. (**a**) Design process of disordered metasurface. (**b**) PL and (**c**) EL spectra of LEDs with and without Ag nanoparticles. (**d**) The output power and voltage of LEDs with and without Ag nanoparticles are a function of current. (**e**) IQE and EQE of LEDs with and without Ag nanoparticles. Reproduced with permission from [17], [*Light: Science & Applications*]; published by Nature Publishing Group, 2021.

In perovskite LEDs (PeLEDs), nanobricks were embedded in electron transport layers to enhance the LEE [23]. This paper presented that the nanopatterned PeLEDs not only improved the LEE, but also showed significant directional emission. The fundamental reason is that due to the diffraction effect of the nanopattern of the medium, more photons fall into the escape cone, resulting in a directional emission pattern. The far-field intensity of PeLEDs with Ag nanopatterns was significantly increased by eight times compared with planar PeLEDs. The optical power loss ratios of planar and nanopatterned PeLEDs at 780 nm wavelength were investigated. The ratios presented that the waveguide mode caused by the large refractive index difference between the active layer and ITO layer is an important factor limiting the LEE.

2.2. Improvement in the LEE by Metasurface Structure in Organic LEDs (OLEDs) or Micro-OLEDs

Metasurfaces are also widely used in OLEDs or Micro-OLEDs to improve the LEE. In 2016, Zhou et al. proposed integrating a speckle image holography (SIH) metasurface at the top to obtain OLEDs with high contrast and high efficiency [24], as shown in Figure 3a. The SIH metasurface, due to its "regional characteristics", shown in Figure 3b, has a wide viewing angle and high contrast directional gain that allows control of the non-interference wavefront generated by the emission layer. Compared with OLEDs with one-dimensional gratings integrated at the top, the efficiency improvement in the SIH metasurface is not affected by wavelength, while the enhanced LEE of the one-dimensional gratings is dependent on wavelength and angle due to resonance (see Figure 3c). The power and EQE improvement in the SIH metasurface are 1.5 times and 1.4 times of one-dimensional gratings, respectively. Next year, by experimenting with three different OLEDs, fluorescent green OLEDs, phosphorescent red OLEDs, and phosphorescent blue OLEDs, the team compared the results of SIH metasurface OLEDs with one-dimensional gratings and flat OLEDs, and found the same experimental results. These results demonstrated that the SIH metasurface improved the LEE and light control independent of material, wavelength, and radiation angle [29].

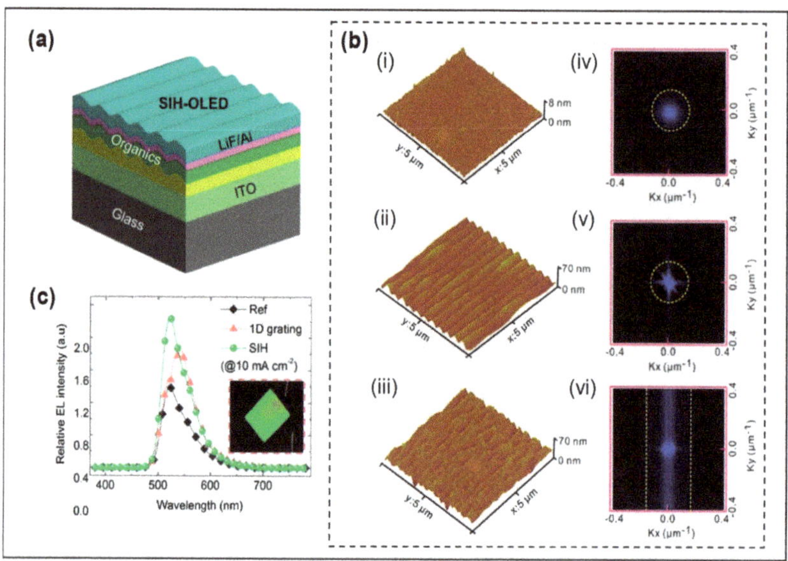

Figure 3. (**a**) Schematic diagram of SIH OLED device structure. (**b**) AFM images of (i) flat surface, (ii) one-dimensional grating patterned surface, and (iii) SIH patterned surface. (iv–vi) The corresponding fast Fourier transform patterns of the AFM images in (i–iii), respectively. (**c**) Relative EL spectrum perpendicular to the direction of the glass substrate. Reproduced with permission from [24], [*ACS Appl. Mater. Interfaces*]; published by American Chemical Society, 2016.

In OLEDs, in addition to the SIH metasurface, supergrating structure is also a common method to improve the LEE of OLEDs [25,26]. Compared with OLEDs without supergratings, the light output intensity of supergrating OLEDs is 4.8 times higher at 510 nm. The OLED structure with supergratings is shown in Figure 4a. The reflected supergratings effectively couple the beam captured in the waveguide mode to enhance the LEE. The team compared the effects of periodic and quasi-periodic supergratings on the luminescence intensity of OLEDs and found that although periodic supergratings are enhanced more, the polarization dependence is very strong. Therefore, periodic supergratings need to be replaced by quasi-periodic supergratings in some specific applications. Kang et al. reported a nanoslot metasurface enhances the LEE of OLEDs [27]. As shown in Figure 4b, the structural parameters of nanoslots include width W and length L1 and L2. The layer cross-section of nanoslot metasurface OLEDs and flat OLEDs are shown in Figure 4c. The addition of nanoslot metasurface can induce surface plasmon (SP) and localized SP mechanisms to enhance the external coupling efficiency and reduce the ambient light reflectance of OLEDs. The change in metasurface layer thickness has an influence on both external coupling efficiency and environmental reflectance; thus, the properties of both should be considered to strike a balance when changing the thickness. The structural parameters of nanoslots have little effect on the external coupling efficiency, but have a great effect on the reflectance. Therefore, a small reflectance can be obtained by adjusting the structural parameters of nanoslots. The performance of the optimized nanoslot metasurface OLEDs is 15% higher than that of the traditional flat OLEDs.

In Micro-OLEDs, Lin et al. proposed using metalens to enhance the LEE, as shown in Figure 5a [28]. Through simulations, it was found that metalens can convert different wave vectors into directly emitted wave vectors in a single interaction with light, thereby improving the LEE of Micro-OLEDs. It is worth noting that metalens does not require multiple interactions with photons, i.e., multiple reflections are not needed. Therefore, metalens exhibits strong optical coupling effects. As shown in Figure 5b,c, when the focal length is between 1.6 and 2.2, Micro-OLEDs with metalens exhibit a significant improvement in EQE compared to traditional Micro-OLEDs. These technologies are expected to boost AR/VR.

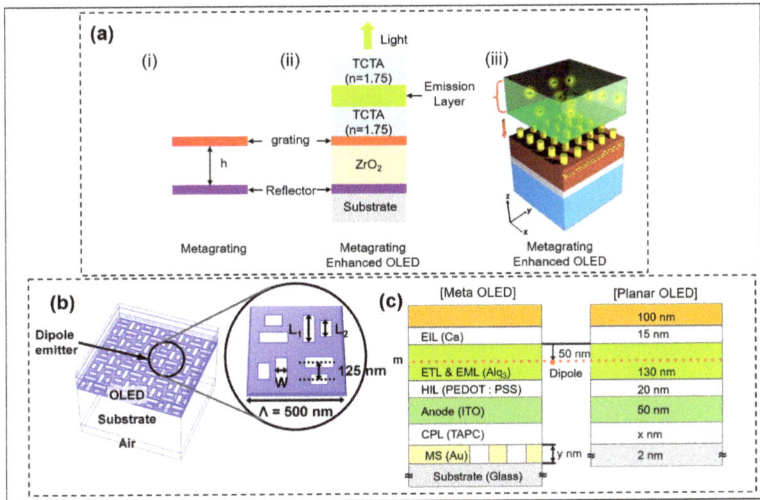

Figure 4. (**a**): (i) Schematic diagram of grating layer structure h height from the bottom mirror. (ii) Layer and (iii) 3D structure diagram of OLEDs integrated with metagrating. Reproduced with permission from [26], [*Applied Physics Letters*]; published by AIP Publishing, 2021. (**b**) 2D layer cross-section of nanoslot metasurface patterned and planar bottom emitting OLEDs. (**c**) 3D structure diagram of the bottom emitting OLEDs integrated nanoslot metasurface. The illustration shows one of the units of the nanoslot metasurface OLEDs. Reproduced with permission from [27], [*Scientific Reports*]; published by Nature Publishing Group, 2021.

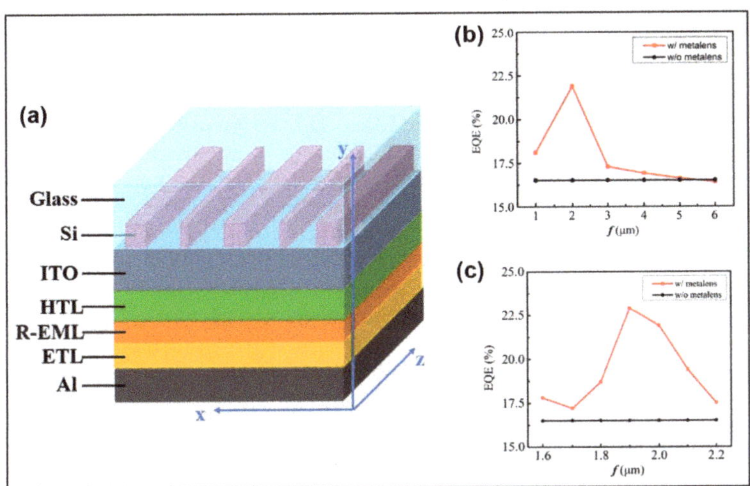

Figure 5. (**a**) A schematic diagram of the structure of Micro-OLEDs with metalens. (**b**) EQE with/without metalens at focal lengths (**b**) of 1 to 6 μm and (**c**) of 1.6 to 2.2 μm. Reproduced with permission from [28], [*Advanced Photonics Research*]; published by Wiley-VCH GmbH, 2021.

3. Improvement in the Emitted Light Collimation

In micro-displays composed of pixel arrays formed by Micro-LEDs, the close proximity of pixels and the Lambertian-shaped emission of Micro-LEDs result in optical crosstalk between adjacent pixels, affecting display clarity. Improving the collimation of Micro-LED light emission is beneficial for reducing optical crosstalk between adjacent pixels and improving display performance. Furthermore, metasurface structures can only modulate light with strong spatial coherence, so it is necessary to improve the collimation of Micro-LEDs in order to be controlled by metasurfaces and generate more functionalities. Currently, methods to improve the collimation of Micro-LEDs include resonant cavities (RC) [31], plasma collimation [32,33], lens collimation [34,35], and light-blocking patterns between pixels [36]. However, methods like using light-blocking patterns between pixels achieve overall collimation of Micro-LEDs rather than collimation of the Micro-LEDs' intrinsic emission. Therefore, such methods cannot achieve metasurface control over Micro-LEDs. Currently, the most mainstream method to achieve metasurface control over the phase of Micro-LED emission is through RC. The recent research progress in improving the collimation of LEDs is summarized in Table 2.

Additionally, the thickness of OLEDs and PeLEDs is only a few hundred nanometers, while Micro-LEDs are a few micrometers. Therefore, the interference effect of the F-P cavity is more obvious in OLEDs and PeLEDs. Moreover, due to the difference in device structure, the emission position of Micro-LEDs is within the multi-quantum wells (MQWs), resulting in different heights of the light sources, which makes the microcavity effect more complex. For some Mcro-LEDs with strong cavity effects, this is mainly because they apply thin-film Micro-LEDs. It should be noted that the design of the top and bottom mirrors of the RC is related to the absorption of the Micro-LED itself [37]. Considering the absorption losses of the Micro-LEDs, high reflectivity mirrors, such as metal mirrors or a large stack DBR, should be used on the non-emitting surface to prevent light leakage. On the emitting surface, mirrors with moderate reflectivity should be used to prevent the excessive back-and-forth reflection of light and minimize absorption losses.

Bottom reflector technology can reflect the Micro-LED active layer emitted light and the top fully reflected light back to the top, increasing the efficiency of the emitted light. Bai et al. proposed the integration of lattice-matched DBR at the bottom of the multi-quantum wells (MQWs), shown in Figure 6a [38]. The bottom DBR consists of 11 pairs of lattice-matched nanoporous (NP) GaN/undoped GaN. NP-GaN was obtained by electrochemical (EC) etching of the n^{++}-GaN layer. The addition of the bottom mirror prevents light from leaking out on the bottom substrate, allowing more light to be reflected and emitted from the top to improve the LEE. Figure 6b presents a planar scanning electron microscopy (SEM) image of a Micro-LED wafer with a diameter of 3.6 μm and a spacing of 2 μm. The team developed a method to manufacture Micro-LEDs using selective overgrowth, which avoided the side wall damage caused by dry etching. A 9% ultra-high EQE was eventually obtained when the working wavelength was 500 nm, and the spectral width of the luminescence was reduced to 25 nm.

Regarding adding the bottom reflector LEDs, adding the top reflector can form the F-P cavity. The cavity makes the LED that is emitting light carry out constructive interference and destructive interference in the cavity, which makes the emitted light more collimated after the resonance selection. Huang et al. introduced the F-P cavity into GaN-based Micro-LEDs to improve collimation [18]. The RC Micro-LED structure is shown in Figure 6c. Both the bottom mirror and the top mirror of the F-P cavity are composed of SiO_2/TiO_2 DBR, but the logarithm is different. In this paper, the reflectivity changes in DBR with different logarithms at 450 nm were studied. Figure 6d shows that the more logarithms there are, the higher the reflectivity. The effect of bottom and top DBR reflectivity on the enhancement factor of light extraction was also studied. This provides an idea for how to set up a suitable cavity structure in Micro-LEDs. As shown in Figure 6e, when the working wavelength is 450 nm, the logarithm of DBR at the bottom is at least 10 pairs to make the reflectivity close to 100%, and the logarithm of DBR at the top is 3 pairs to make the reflectivity 65%. The F-P cavity constructed in this way can ensure light extraction when the wavelength satisfying the microcavity effect is resonated multiple times, and multiple resonant selections also make the FWHM of the emission spectrum narrow. GaN-based RC Micro-LEDs with a divergence angle of 78.7° and spectral width of 6.8 nm were made. More emitted light along the normal direction also gives the device great potential in AR applications, such as AR glasses and AR Head Up Display. Because of the resonance selection of the resonator, the output spectrum is narrowed, and the emission is directed.

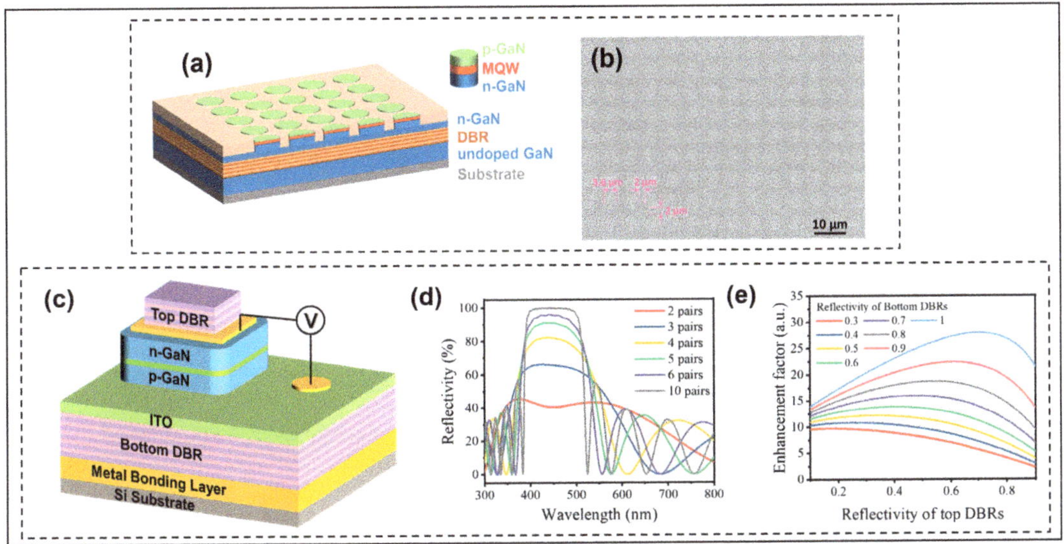

Figure 6. (**a**) Schematic diagram of the incorporation of Micro-LEDs and DBR. (**b**) Planar SEM image of a Micro-LED wafer. Reproduced with permission from [38], [*ACS Nano*]; published by American Chemical Society, 2020. (**c**) Schematic diagram of RC Micro-LED structure. (**d**) The effective reflectivity for different pairs of SiO_2/TiO_2 DBRs. (**e**) Relationship between top and bottom DBR reflectivity and light extraction enhancement factor. Reproduced with permission from [18], [*Applied Physics Letters*]; published by AIP Publishing, 2022.

In addition to applications in Micro-LEDs, the resonator can also enhance the LEE in OLEDs, PeLEDs, and quantum dot (QD) LEDs. In 2020, Joo et al. implemented an OLED display technology of over 10,000 pixels per inch by introducing an F-P cavity. The F-P cavity consists of a metasurface Ag mirror as a bottom mirror and an Ag electrode as a top mirror [39], shown in Figure 7a. The introduction of an F-P cavity reduces the divergence angle of OLEDs. The results in Figure 7b,c show that the luminous intensity and IQE of OLEDs with a metasurface F-P cavity are significantly higher than that of white OLEDs with a color filter. The three colors blue, green, and red represent blue light, green light, and red light, respectively. In addition to the color filter itself will reduce the OLED luminous intensity, the F-P cavity will carry out constructive interference on the light wave meeting the resonance condition of the microcavity to increase the intensity of the emitted light. In addition, by changing the diameter and density of nanocolumns in the bottom Ag mirror, different wavelengths of light can be enhanced by resonance to achieve color changes. In this way, OLED displays with smaller pixels can be achieved. Recently, Liang et al. introduced resonators into PeLEDs and developed a resonance enhancer cycling method [40], enabling the color of the emitted light to reach unprecedented purity after resonance selection, which provides a wide color gamut for the development of next-generation displays.

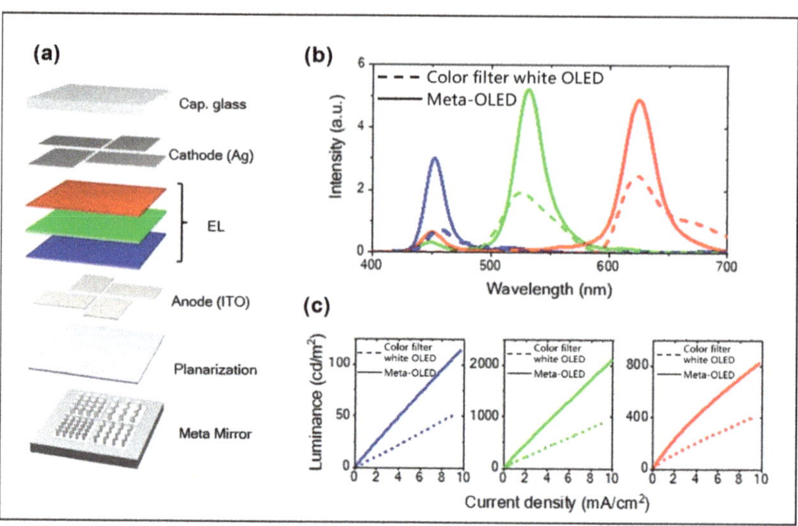

Figure 7. (**a**) Schematic diagram of meta-OLED design for metasurface mirror. (**b**) Electroluminescence (EL) spectrum and (**c**) luminance as a function of current density. Reproduced with permission from [39], [*Science*]; published by American Association for the Advancement of Science, 2020.

Table 2. Summary of research on improving the emitted light collimation of LEDs.

Research Objective	LED Type	Wavelength	Optimized Structure Model	Simulation or Experiment	Ref.
Improve LEE and enhance spectral narrowing	Micro-LED	450~660 nm	Bottom NP GaN/undoped GaN DBR reflectors	Experiment	[38]
Improve LEE and enhance spectral narrowing	Micro-LED	440~460 nm	SiO_2/TiO_2 DBR reflectors F-P cavity	Simulation and experiment	[18]
Improve LEE and reduce pixel size	OLED	370~700 nm	Metasurface Ag and flat Ag mirrors F-P cavity	Simulation and experiment	[39]
Enhance spectral narrowing to improve color purity	PeLED	450~650 nm	Au mirror and DBR reflectors F-P cavity	Experiment	[40]

4. Control of the Deflection of Light Angle

The metasurface structure can not only improve the LEE of Micro-LEDs, but also accurately control the phase of the wavefront to realize the control of the emitted light angle of Micro-LEDs. The nanocolumn array is integrated on the top surface of Micro-LEDs, which is composed of nanocolumns with different diameters in one cycle to achieve 0–2π phase coverage. The phase of nanocolumns can be changed by changing the height of nanocolumns and the equivalent refractive index, which can be achieved by changing the diameter of nanocolumns. In the finite-difference time-domain method, the relationship between the diameter and phase of nanocolumns is obtained by scanning nanocolumns of fixed height and different diameters. The phase coverage of 2π is realized by selecting suitable nanocolumns with different diameters and adjusting different periods to achieve different emission angles. However, only coherent wavefronts can be controlled by metasurface structures, while Micro-LED emission meets the Lambertian-shaped emission and has poor coherence. Therefore, a resonator structure is added to Micro-LEDs to enhance the coherence of the emitted light and realize the metasurface structure to control the angle of the emitted light.

In 2018, Liu et al. demonstrated for the first time the control of metasurface structure on light emitted by LEDs with a wavelength of 460 nm through simulation [13]. This paper compared the far-field patterns of LEDs with directly added TiO_2 nanocolumn arrays and resonant cavity LEDs (RCLEDs) with TiO_2 nanocolumn arrays added to their surfaces. RCLEDs could achieve a 20° angular deflection by controlling the phase with metasurface structures, while LEDs could not. The result indicates that the RC structure can collimate the Lambertian-shaped emission of LEDs, approximating a plane wave, and enhance the spatial coherence of the emitted light from Micro-LEDs. Two years later, the team added the metasurface RCLED design to the GaP LEDs [41], proposing that as long as the reflectivity of the metasurface is low, resonators and metasurface can be designed separately and independently, showing the flexibility and universality of the metasurface design. In the GaP metasurface RCLEDs, the resonator consists of bottom Au mirrors and top DBR mirrors, with top nanocolumns selected with high-index Si materials. However, Si material has high absorption in the visible band, which will reduce the light output efficiency, so TiO_2 material with a high refractive index can be chosen to replace it. In this paper, the 30° deflection of the emitted light beam with a wavelength of 620 nm was obtained (see Figure 8d). The far-field patterns of the experimental results correspond well with the simulation results. Figure 8 shows the full process of angular control through the integration of metasurface structures and RCLEDs.

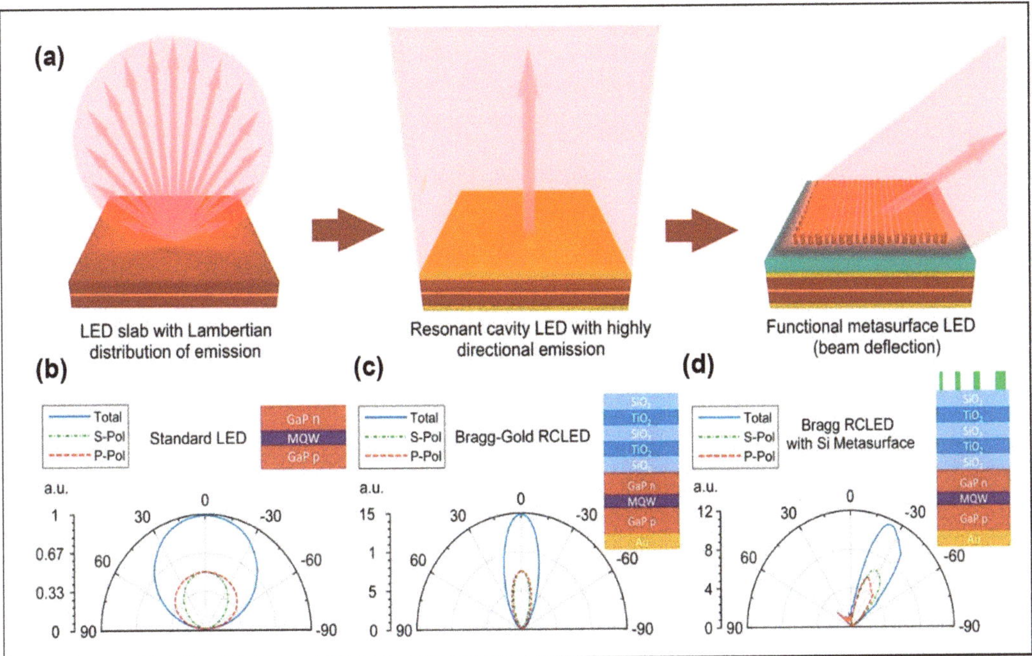

Figure 8. (a) Schematic diagram of metasurface control of LED-emitted light. The far-field diagram of (b) GaP LEDs, (c) Hybrid Bragg–gold RCLEDs, and (d) Hybrid RCLEDs with the integrated metasurface. Reproduced with permission from [41], [*Laser Photonics Reviews*]; published by Wiley-VCH GmbH, 2020.

In recent years, the demand for near-eye display devices has been increasing, and Micro-LEDs that can be emitted in unidirectional directions are undoubtedly one of the most ideal light sources, as unidirectional emissions can reduce optical crosstalk and provide a better visual experience. Huang et al. proposed a metasurface RC Micro-LEDs with unidirectional emission [19]. Different diameters of nanocolumns can produce different phase changes, which can satisfy the phase coverage of 0–2π within a period. The nanocolumns with different diameters are selected to form different periods to achieve unidirectional emission of light at different angles. The structure diagram and control phase schematic diagram are shown in Figure 9a,b, respectively. Controllable unidirectional emission Micro-LEDs can prevent light leakage. In a 3D display, unidirectional emission of light to the left eye and the right eye can reduce pixel crosstalk and improve the 3D display effect, shown in Figure 9c. In addition, Micro-LEDs composed of a pixel with a variety of emission directions can produce a wider viewing angle and more 3D viewing points (see Figure 9d), so as to have a better 3D display experience, providing more impetus for the future development of near-eye 3D Micro-LED displays. The team later made a 3D display using Micro-LEDs emitted unidirectionally and compared it with a traditional 3D display [42]. Through comparison, the results showed that unidirectional emission Micro-LED pixels can effectively reduce optical crosstalk and improve the display effect.

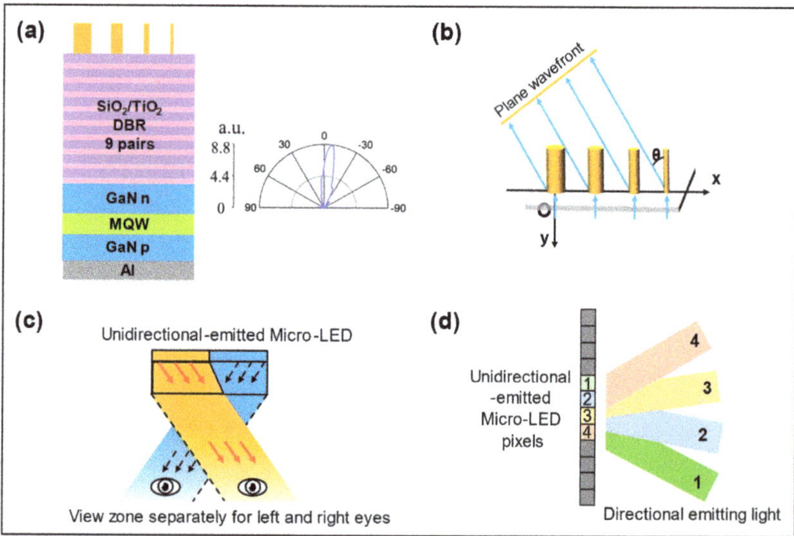

Figure 9. (**a**) Device structure and far-field diagram of Micro-LEDs with the beam deflecting metasurface. (**b**) Working schematic of the beam deflection supercell. (**c**) Unidirectional emission Micro-LEDs for naked-eye 3D display. (**d**) Multi-view naked-eye 3D display with unidirectional emission Micro-LED pixels. Reprinted/Adapted with permission from [19] © The Optical Society.

The control of the angle of emitted light by metasurface structure has also been developed in QD LEDs, which brings a new scheme for optimizing QD LEDs. Park et al. showed the control of metasurface structures on the angles of light emitted by the colloidal quantum dot (CQD) RCLEDs [43]. The CQD RCLED structure is shown in Figure 10a with five pairs of DBR mirrors on both the bottom and top to form a resonator, with the top integrating TiO_2 nanocolumns to allow angular deflection. In Figure 10b, the relationship between the diameter and phase (red lines) of TiO_2 nanocolumns with heights of 700 nm and 300 nm is shown, respectively. The results present that the phase coverage of 0–2π cannot be achieved by the short nanocolumns, but using patterned TiO_2 nanocolumns with a high aspect ratio is difficult. Therefore, the team selected TiO_2 nanocolumns with low aspect ratios to control the angle of light by adjusting the period and phase within the period, thus reducing the difficulty of the manufacturing process. The black line represents the relationship between diameter and transmittance, and it can be observed that the high TiO_2 nanorods experience a sharp drop at a diameter of 250 nm. The final simulated and experimentally measured deflection angles in this paper are both 20°, with only a 4 nm deviation in peak wavelength. Therefore, the experimental results highly match the simulation. Huang et al. conducted a simulation study on the exit angle control of QD LEDs [44]. Comparing the emission light direction diagrams of the Gaussian beam and electric dipole light sources in the RCLEDs formed by the bottom flattened Ag mirror and top DBR mirror, they found that the RC of the flattened Ag mirror and DBR mirror failed to collimate the QD LED emission of light, resulting in unrealized angle control. The team proposed that the circular patterned bottom Ag mirror and the top DBR mirror formed an RC to realize the emitted light collimation, and the different deflection angles were controlled by selecting nanocolumns of different diameters to form different cycles, shown in Figure 10c. This method provides a scheme for the manipulation of a single QD, and presents the structural design differences between QD LEDs and Micro-LEDs in angular deflection.

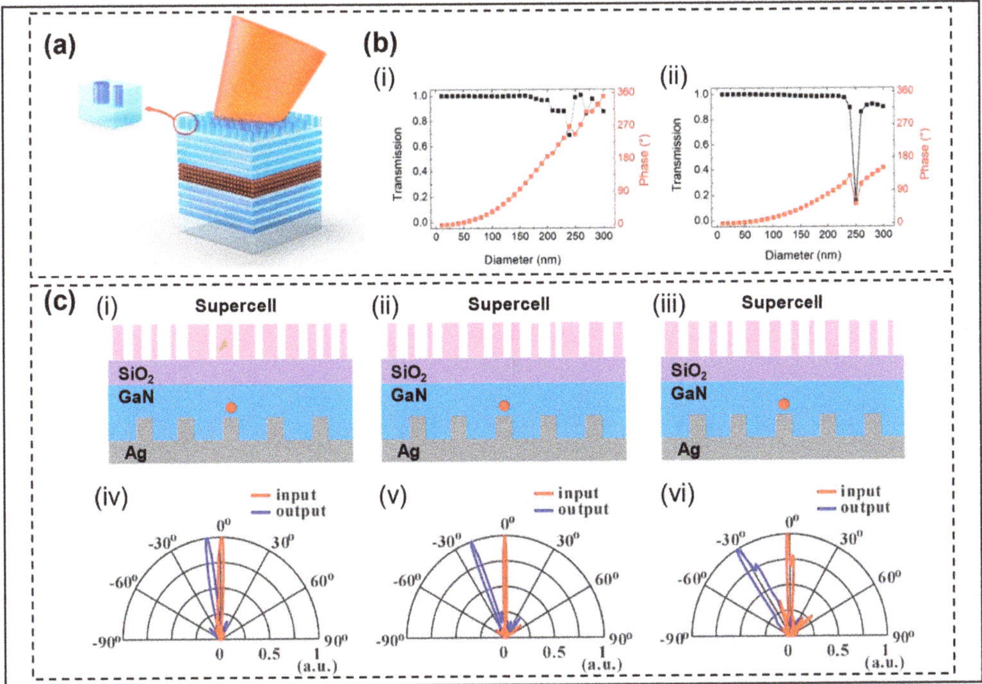

Figure 10. (**a**) Schematic of CQD RCLEDs with TiO_2 metasurface. (**b**) Calculated look-up tables of a TiO_2 nanocolumn with thicknesses of (i) 700 nm and (ii) 300 nm. Reproduced with permission from [43], [*Nanophotonics*]; published by De Gruyter, 2020. (**c**) (i–iii) Schematic structures of GaN-based QD LEDs with an Ag grating and a phase-gradient metasurface. Polar plots for the input beam and output beam of the metasurface. (iv–vi) are the cases for 10°, 20°, and 30° angle deflection, respectively. Reprinted/Adapted with permission from [44] © The Optical Society.

In Table 3, the research on metasurface structures controlling the direction of light emitted by Micro-LEDs is summarized. Recent research findings are mostly simulation results, as the manufacturing process of multi-layer DBR mirrors and metasurface is relatively complex, and the design and fabrication of metasurface structures still face many challenges and high costs. Therefore, there might still be a long way before their application in displays.

Table 3. Summary of research on metasurface structure to control LED emission angle.

Research Objective	LED Type	Wavelength	Optimized Structure Model	Simulation or Experiment	Ref.
Achieve light directional emission at the expected angle	LED	460 nm	Top TiO_2 nanocolumns, DBR reflectors, and Al mirror cavity	Simulation	[13]
Achieve light directional emission at the expected angle	GaP LED	615~640 nm	Top Si nanocolumns, DBR reflectors, and Au mirror cavity	Simulation and experiment	[41]
Achieve light directional emission at the expected angle	CQD LED	580~620 nm	Top TiO_2 nanocolumns and DBR reflectors cavity	Simulation and experiment	[43]
Achieve light directional emission at the expected angle	QD LED	520 nm	Top TiO_2 nanocolumns and bottom circular patterned Ag grating	Simulation	[44]
Realize Micro-LED unidirectional emission	Micro-LED	445 nm	Top TiO_2 nanocolumns, DBR reflectors, and Al mirror cavity	Simulation	[19]
Reduce pixel crosstalk by unidirectional emission Micro-LED	Micro-LED	460 nm	Top TiO_2 nanocolumns, DBR reflectors, and Al mirror cavity	Simulation	[42]

5. Control of Near-Eye Polarization

Polarized LEDs play an important role in 3D displays, providing linearly polarized (LP) and circularly polarized (CP) light. Currently, the most common way for LEDs to produce LP light is to use a grating structure. By changing the period, duty cycle, and thickness of the grating, the transmission of transverse magnetic (TM) wave and the reflection of transverse electric (TE) wave are controlled to realize the emission of LP light. In 2010, Zhang et al. further optimized the grating structure and proposed that adding a dielectric transition layer with a lower refractive index than the GaN layer between the GaN layer and metal grating could improve the polarization characteristics of LEDs, that is, the transmission of TM wave and ER [45]. Ma et al. added Al grating on a sapphire substrate to realize LEDs with LP light emitted from the back, and the polarization degree could reach 0.96 [46]. Huang et al. directly integrated subwavelength metal gratings on the p-GaN surface and achieved a high ER of 14.17 dB by optimizing the grating period, thickness, and width [47]. Although LP light is realized through a grating structure, the energy of the TE wave is lost, and the luminous efficiency is reduced. Zhang et al. proposed the coupling effect between MQWs and metal gratings to generate surface plasma to improve the radiation recombination rate and polarization degree [48]. The team's approach was to etch the p-GaN layer into a grating structure and coat it with a layer of Al grating to achieve linear polarization, as shown in Figure 11a. This method improves the luminous efficiency, but it does not fundamentally solve the TE wave loss.

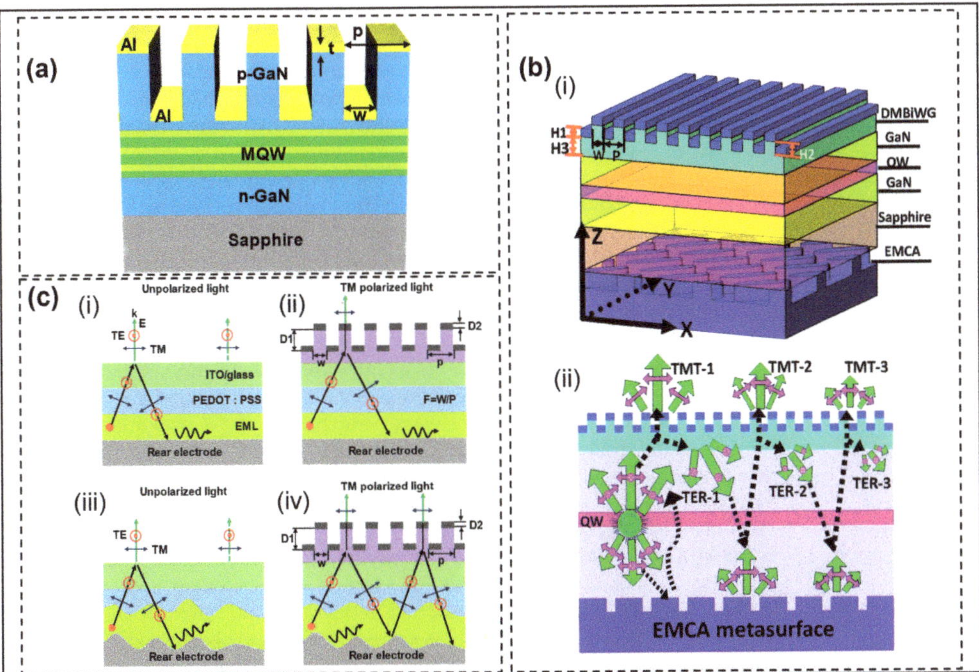

Figure 11. (**a**) Schematic diagram of the polarized LED structure. Reproduced with permission from [48], [*ACS Photonics*]; published by American Chemical Society, 2016. (**b**): (i) Structure diagram of InGaN/GaN LEDs with integrated metasurface and metal/dielectric grating structure. (ii) Propagation and polarization conversion processes of both TM and TE polarized components in the integrated structure. Reproduced with permission from [16], [*Nanoscale*]; published by Royal Society of Chemistry, 2017. (**c**) Schematic cross-section of device structures with (i) flat OLEDs, (ii) polarized OLEDs with D/M nanograting, (iii) OLEDs with holographic metasurface of nanospeckle image, and (iv) OLEDs integrated with D/M nanograting and holographic metasurface of nanospeckle image. Reproduced with permission from [49], [*Laser Photonics Reviews*]; published by Wiley-VCH GmbH, 2020.

To solve the TE wave energy loss, studies have focused on the application of metasurface structures in polarized LEDs. Wang et al. presented an LED with integrated metasurface structures and dielectric/metal (D/M) double-layer grating structures [16], shown in Figure 11b. The top gratings are used to transmit the TM wave and reflect the TE wave to generate LP light. The bottom metasurface nanocolumns act as half-wave plates to convert TE waves into TM waves, thus reducing the loss of TE waves. Both the luminous efficiency and ER are improved. This method provides a new idea for high-efficiency LP LEDs. Zhou et al. later applied the idea of the interaction of metasurface and gratings to OLEDs and proposed the integrated D/M nanograting structure at the top and holographic metasurface structures of the nanospeckle image at the bottom (see Figure 11c(iv)), realizing a white OLEDs with high-efficiency linear polarization with ultra-high polarization ratio [49]. In Figure 11c, the comparison of the four structure diagrams shows that the holographic metasurface not only converts TE waves into TM waves to improve emission efficiency, but also changes the emission direction of reflected light so that more light is emitted in the escape cone.

In 3D displays, LP light requires that the left and right eyes remain at the same level in order to reduce crosstalk between the left and right images, which makes for a very bad viewing experience. CP light overcomes this difficulty, reduces optical crosstalk, and improves the 3D display effect. Gao et al. designed CP Micro-LEDs using metasurface and grating structures [15], and the structure diagram is shown in Figure 12a(i). Regarding RC Micro-LEDs, the Al grating structure is integrated at the bottom to transmit TM waves, reflect TE waves, and treat TE waves as emission light, which also leads to loss of TM wave energy and a decrease in luminescence efficiency. The nanobricks integrated on the top are used as a quarter-wave plate to convert LP light to CP light, which provides a new idea for the design of CP Micro-LEDs. The TE reflectivity (RTE) and TM reflectivity (RTM) are changed by adjusting the period and duty cycle of the bottom grating to obtain high ER, as shown in Figure 12a(ii–iv). As a tool to flexibly modulate the polarization of light, the wave plate can change LP light into CP light, but the problems of high efficiency and tunable phase delay need to be solved. In OLEDs, Wu et al., using liquid crystal polymer as a substrate, proposed to prepare ultra-thin, flexible, foldable, and stretchable wave plates by a water-soluble transfer method [50]. The wave plate can achieve a delay effect at any wavelength of the visible band and has a high transmittance; most of the light transmittance is above 95%. Therefore, the use of this wave plate in flexible OLEDs will not lose the emitted light. In addition to using a wave plate, Jia et al. developed micro-cavity CP OLEDs through chiral light emitting body [51], shown in Figure 12b, which not only provided a new idea for CP OLEDs, but also overcame the problem of a chiral light emitting body reducing EL. Chiral emitter OLEDs have two Ag layers at the top and bottom as two metal mirrors and electrodes of OLEDs. A thin layer of two-dimensional organic single crystal is embedded between Ag layers. The advantage of an organic single crystal is that the anisotropy of its refractive index will lead to a birefringent microcavity, which is conducive to the occurrence of the Rashba–Dresselhaus (RD) effect and changes the linear polarization into circular polarization. Finally, the study of metasurface structures on polarization control of LEDs is summarized in Table 4.

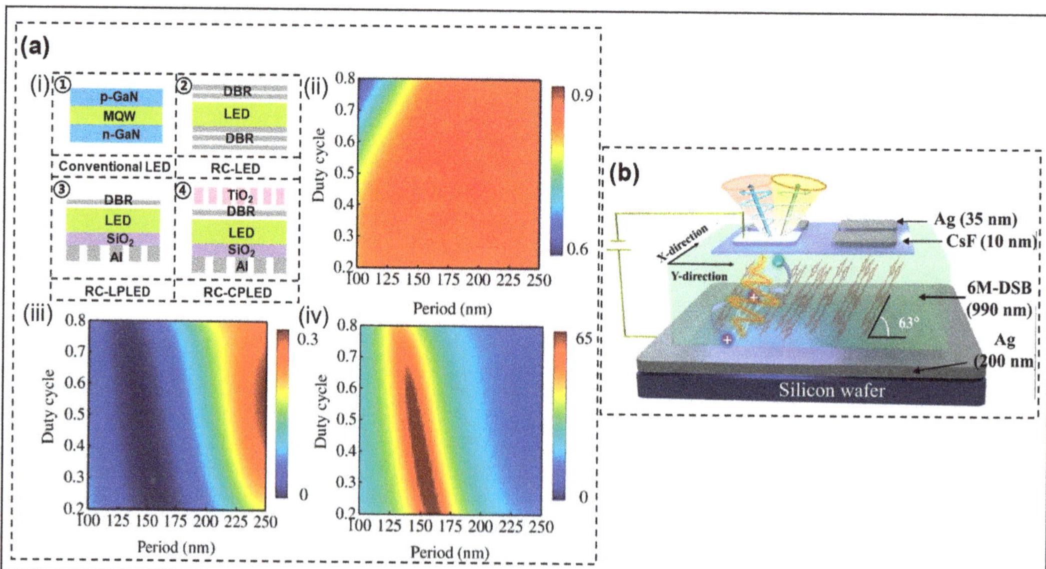

Figure 12. (**a**): (i) Cross-sectional views for different types of Micro-LEDs. (ii) RTE, (iii) RTM, and (iv) ER of Al nanograting with different periods and duty cycles calculated from numerical simulation. Reprinted/Adapted with permission from [15] © The Optical Society. (**b**) Schematic diagram of the microcavity CP OLED structure. Reproduced with permission from [51], [*Nature Communications*]; published by Nature Publishing Group, 2023.

Table 4. Summary of research on metasurface structure to control LED polarization.

Research Objective	LED Type	Wavelength	ER or Polarization Degree	Optimized Structure Model	Simulation or Experiment	Ref.
Improve extinction ratio of LP light	LED	470 nm	60 dB	Dielectric transition grating	Simulation	[45]
Improve polarization degree of LP light	LED	445~470 nm	0.96	Wire-grid polarizer on sapphire	Simulation and experiment	[46]
Improve extinction ratio of LP light	Micro-LED	400~480 nm	14.17 dB	Subwavelength metal grating	Simulation and experiment	[47]
Improve polarization luminous efficiency and polarization degree of LP light	LED	500~540 nm	0.54	P-GaN and Al grating	Simulation and experiment	[48]
Improve polarization luminous efficiency and extinction ratio of LP light	LED	500~560 nm	20 dB	Top dielectric/metal grating and bottom elliptical metal nanocolumns	Simulation and experiment	[16]
Improve polarization luminous efficiency and extinction ratio of LP light	OLED	450~650 nm	17.8 dB	Top dielectric/metal grating and bottom holographic metasurface	Simulation and experiment	[49]
Obtain CP light	Micro-LED	450 nm	38 dB	Top TiO_2 nanobricks and bottom Al grating	Simulation	[15]
Improve the luminous efficiency of CP light	OLED	450~700 nm	/	Waveplates based on de-ionized water immersion transfer method	Experiment	[50]
Improve the luminous efficiency of CP light	OLED	450~550 nm	/	Embedding a thin two-dimensional organic single crystal	Simulation and experiment	[51]

6. Conclusions

In this paper, the applications of metasurface structures integrated with Micro-LEDs are reviewed from three aspects: to improve the LEE, to achieve directional emission, and to realize LP light and CP light. The reason for increasing the LEE is that the addition of metasurface structures will reduce the total reflection consumption of LEDs at the air-contact surface, so that more light is emitted in the escape cone. Control of Lambertian-shaped emission LED light by metasurface structures requires the use of RC to increase spatial coherence. Meanwhile, the realization of LP and CP light further accelerates the use of Micro-LEDs in 3D near-eye displays. Metasurface structures have also been widely used in OLEDs, QD LEDs, and PeLEDs, and optimized results have been obtained to improve light extraction efficiency and directional emission. In summary, the integration of metasurface structures provides more possibilities for the structural design and performance optimization of Micro-LEDs.

Funding: This research was funded in part by the National Natural Science Foundation of China under Grant 62204150, and the Science and Technology Commission of Shanghai Municipality Program under Grant 21511101302 and Grant 20010500100.

Data Availability Statement: Not applicable.

Conflicts of Interest: The authors declare no conflict of interest.

References

1. Lee, H.E.; Shin, J.H.; Park, J.H.; Hong, S.K.; Park, S.H.; Lee, S.H.; Lee, J.H.; Kang, S.S.; Lee, K.J. Micro Light-Emitting Diodes for Display and Flexible Biomedical Applications. *Adv. Funct. Mater.* **2019**, *29*, 1808075. [CrossRef]
2. Wang, Z.; Shan, X.; Cui, X.; Tian, P. Characteristics and techniques of GaN-based micro-LEDs for application in next-generation display. *J. Semicond.* **2020**, *41*, 041606. [CrossRef]
3. Liu, Z.; Lin, C.; Hyun, B.; Sher, C.; Lv, Z.; Luo, B.; Jiang, F.; Wu, T.; Ho, C.; Kuo, H.; et al. Micro-light-emitting diodes with quantum dots in display technology. *Light Sci. Appl.* **2020**, *9*, 83. [CrossRef] [PubMed]
4. Wu, T.; Sher, C.; Lin, Y.; Lee, C.; Liang, S.; Lu, Y.; Chen, S.H.; Guo, W.; Kuo, H.; Chen, Z. Mini-LED and Micro-LED: Promising Candidates for the Next Generation Display Technology. *Appl. Sci.* **2018**, *8*, 1557. [CrossRef]
5. Huang, Y.; Hsiang, E.; Deng, M.; Wu, S. Mini-LED, Micro-LED and OLED displays: Present status and future perspectives. *Light Sci. Appl.* **2020**, *9*, 105. [CrossRef]
6. Anwar, A.R.; Sajjad, M.T.; Johar, M.A.; Hernández-Gutiérrez, C.A.; Usman, M.; Łepkowski, S.P. Recent Progress in Micro-LED-Based Display Technologies. *Laser Photonics Rev.* **2022**, *16*, 2100427. [CrossRef]
7. Wong, M.S.; Nakamura, S.; DenBaars, S.P. Review-Progress in High Performance III-Nitride Micro-Light-Emitting Diodes. *ECS J. Solid State Sci. Technol.* **2019**, *9*, 015012. [CrossRef]
8. Zhuang, Z.; Iida, D.; Ohkawa, K. InGaN-based red light-emitting diodes: From traditional to micro-LEDs. *Jpn. J. Appl. Phys.* **2021**, *61*, SA0809. [CrossRef]
9. Chang, S.; Guo, X.; Ni, X. Optical metasurfaces: Progress and applications. *Annu. Rev. Mater. Res.* **2018**, *48*, 279–302. [CrossRef]
10. Yang, J.; Gurung, S.; Bej, S.; Ni, P.; Lee, H.W.H. Active optical metasurfaces: Comprehensive review on physics, mechanisms, and prospective applications. *Rep. Prog. Phys.* **2022**, *85*, 036101. [CrossRef]
11. Du, K.; Barkaoui, H.; Zhang, X.; Jin, L.; Song, Q.; Xiao, S. Optical metasurfaces towards multifunctionality and tunability. *Nanophotonics* **2022**, *11*, 1761–1781. [CrossRef]
12. Su, V.; Chu, C.H.; Sun, G.; Tsai, D.P. Advances in optical metasurface: Fabrication and applications. *Opt. Express* **2018**, *26*, 13148–13182. [CrossRef]
13. Liu, Z.; Khaidarov, E.; Akimov, Y.; Paniagua-Domínguez, R.; Sun, S.; Bai, P.; Png, C.E.; Demir, H.V.; Kuznetsov, H.I. Using metasurfaces to control random light emission. In Proceedings of the 2018 Conference on Lasers and Electro-Optics Pacific Rim, Hong Kong, China, 29 July–3 August 2018.
14. Parbrook, P.J.; Corbet, B.; Han, J.; Seong, T.; Amano, H. Micro-Light Emitting Diode: From Chips to Applications. *Laser Photonics Rev.* **2021**, *15*, 2000133. [CrossRef]
15. Gao, X.; Xu, Y.; Huang, J.; Wang, L. Circularly polarized light emission from a GaN micro-LED integrated with functional metasurfaces for 3D display. *Opt. Lett.* **2021**, *46*, 2666–2669. [CrossRef]
16. Wang, M.; Xu, F.; Lin, Y.; Cao, B.; Chen, L.; Wang, C.; Wang, J.; Xu, K. Metasurface integrated high energy efficient and high linearly polarized InGaN/GaN light emitting diode. *Nanoscale* **2017**, *9*, 9104–9111. [CrossRef]
17. Mao, P.; Liu, C.; Li, X.; Liu, M.; Chen, Q.; Han, M.; Maier, S.A.; Sargent, E.H.; Zhang, S. Single-step-fabricated disordered metasurfaces for enhanced light extraction from LEDs. *Light Sci. Appl.* **2021**, *10*, 180. [CrossRef]
18. Huang, J.; Tang, M.; Zhou, B.; Liu, Z.; Yi, X.; Wang, J.; Li, J.; Pan, A.; Wang, L. GaN-based resonant cavity micro-LEDs for AR application. *Appl. Phys. Lett.* **2022**, *121*, 201104. [CrossRef]
19. Huang, J.; Hu, Z.; Gao, X.; Xu, Y.; Wang, L. Unidirectional-emitting GaN-based micro-LED for 3D display. *Opt. Lett.* **2021**, *6*, 3476–3479. [CrossRef]
20. Chen, Z.; Yan, S.; Danesh, C. MicroLED technologies and applications: Characteristics, fabrication, progress, and challenges. *J. Phys. D Appl. Phys.* **2021**, *54*, 123001. [CrossRef]
21. Lee, T.; Chen, L.; Lo, Y.; Swayamprabha, S.S.; Kumar, A.; Huang, Y.; Chen, S.; Zan, H.; Chen, F.; Horng, R.; et al. Technology and applications of micro-LEDs: Their characteristics, fabrication, advancement, and challenges. *ACS Photonics* **2022**, *9*, 2905–2930. [CrossRef]
22. Wang, H.; Wang, L.; Sun, J.; Guo, T.; Chen, E.; Zhou, X.; Zhang, Y.; Yan, Q. Role of surface microstructure and shape on light extraction efficiency enhancement of GaN micro-LEDs: A numerical simulation study. *Displays* **2022**, *73*, 102172. [CrossRef]
23. Ci, Q.; Ren, X.; Yan, Y.; Ren, H.; Niu, K.; Sun, G.; Huang, Z.; Wu, X. The Influence of the Emission Source on Outcoupling and Directivity of Patterned Perovskite Light-Emitting Diode. *IEEE Photonics J.* **2021**, *13*, 8200405. [CrossRef]
24. Zhou, L.; Ou, Q.; Shen, S.; Zhou, Y.; Fan, Y.; Zhang, J.; Tang, J. Tailoring directive gain for high-contrast, wide-viewing-angle organic light-emitting diodes using speckle image holography metasurfaces. *ACS Appl. Mater. Interfaces* **2016**, *8*, 22402–22409. [CrossRef] [PubMed]
25. Xu, X. Fabrication and Integration of Metasurfaces and Metagratings into Organic Photodetectors and Light Emitters. Ph.D. Thesis, The University of Texas at Austin, Austin, TX, USA, 2019.
26. Xu, X.; Kwon, H.; Finch, S.; Lee, J.Y.; Nordin, L.; Wasserman, D.; Alù, A.; Dodabalapur, A. Reflecting metagrating-enhanced thin-film organic light emitting devices. *Appl. Phys. Lett.* **2021**, *118*, 053302. [CrossRef]
27. Kang, K.; Im, S.; Lee, C.; Kim, J.; Kim, D. Nanoslot metasurface design and characterization for enhanced organic light-emitting diodes. *Sci. Rep.* **2021**, *11*, 9232. [CrossRef]
28. Lin, J.G.; Sun, Q.; Feng, W.B.; Guo, S.M.; Liu, Z.h.; Liang, H.W.; Li, J.T. Enhancing the Light Extraction Efficiency in Micro-Organic Light-Emitting Diodes with Metalens. *Adv. Photonics Res.* **2021**, *2*, 2000145. [CrossRef]

29. Zhou, L.; Wang, Q.; Ou, Q.; Zhu, Y.; Lin, Y.; Fan, Y.; Wei, H. Speckle image holography modulated full-color organic light-emitting diodes with high efficiency and engineered emission profile. *Org. Electron.* **2017**, *42*, 13–20. [CrossRef]
30. Yue, Q.Y.; Li, K.; Kong, F.; Zhao, J.; Liu, M. Analysis on the effect of amorphous photonic crystals on light extraction efficiency enhancement for GaN-based thin-film-flip-chip light-emitting diodes. *Opt. Commun.* **2016**, *367*, 72–79. [CrossRef]
31. Zhou, L.M.; Ren, B.C.; Zheng, Z.W.; Ying, L.Y.; Long, H.; Zhang, B.P. Fabrication and Characterization of GaN-Based Resonant-Cavity Light-Emitting Diodes with Dielectric and Metal Mirrors. *ECS. J. Solid State Sci. Technol.* **2018**, *7*, 34–37. [CrossRef]
32. Moreno, E.; Garcia-Vidal, F.J.; Martin-Moreno, L. Enhanced transmission and beaming of light via photonic crystal surface modes. *Phys. Rev. B* **2004**, *69*, 121402. [CrossRef]
33. DiMaria, J.; Dimakis, E.; Moustakas, T.D.; Paiella, R. Plasmonic Collimation and Beaming from LED Active Materials. In Proceedings of the CLEO, San Jose, CA, USA, 9–14 June 2013.
34. Joo, J.Y.; Lee, S.K. Miniaturized TIR Fresnel Lens for Miniature Optical LED Applications. *Int. J. Precis. Eng. Manuf.* **2009**, *10*, 137–140. [CrossRef]
35. Chen, J.J.; Wang, T.Y.; Huang, K.L.; Liu, T.S.; Tsai, M.D.; Lin, C.T. Freeform lens design for LED collimating illumination. *Opt. Express* **2012**, *20*, 10984–10995. [CrossRef]
36. Chung, K.N.; Sui, J.Y.; Sui, J.Y.; Demory, B.; Teng, C.H.; Ku, P.C. Monolithic integration of individually addressable light-emitting diode color pixels. *Appl. Phys. Lett.* **2017**, *110*, 111103. [CrossRef]
37. Delbeke, D.; Bockstaele, R.; Bienstman, P.; Baets, R.; Benisty, H. High-Efficiency Semiconductor Resonant-Cavity Light-Emitting Diodes: A Review. *IEEE J. Sel. Top. Quantum Electron.* **2002**, *8*, 189–206. [CrossRef]
38. Bai, J.; Cai, Y.; Feng, P.; Fletcher, P.; Zhu, C.; Tian, Y.; Wang, T. Ultrasmall, ultracompact and ultrahigh efficient InGaN micro light emitting diodes (μLEDs) with narrow spectral line width. *ACS Nano* **2020**, *14*, 6906–6911. [CrossRef]
39. Joo, W.; Kyoung, J.; Esfandyarpour, M.; Lee, S.; Koo, H.; Song, S.; Kwon, Y.; Song, S.H.; Bae, J.C.; Jo, A.; et al. Metasurface-driven OLED displays beyond 10,000 pixels per inch. *Science* **2020**, *370*, 459–463. [CrossRef]
40. Liang, J.; Du, Y.; Wang, K.; Ren, A.; Dong, X.; Zhang, C.; Tang, J.; Yan, Y.; Zhao, Y.S. Ultrahigh Color Rendering in RGB Perovskite Micro-Light-Emitting Diode Arrays with Resonance-Enhanced Photon Recycling for Next Generation Displays. *Adv. Opt. Mater.* **2022**, *10*, 2101642. [CrossRef]
41. Khaidarov, E.; Liu, Z.; Paniagua-Domínguez, R.; Ha, S.T.; Valuckas, V.; Liang, X.; Akimov, Y.; Bai, P.; Png, C.E.; Demir, H.V.; et al. Control of LED emission with functional dielectric metasurfaces. *Laser Photonics Rev.* **2020**, *14*, 1900235. [CrossRef]
42. Xu, Y.; Cui, J.; Hu, Z.; Gao, X.; Wang, L. Pixel crosstalk in naked-eye micro-LED 3D display. *Appl. Opt.* **2021**, *60*, 5977–5983. [CrossRef]
43. Park, Y.; Kim, H.; Lee, J.; Ko, W.; Bae, K.; Cho, K. Direction control of colloidal quantum dot emission using dielectric metasurfaces. *Nanophotonics* **2020**, *9*, 1023–1030. [CrossRef]
44. Huang, H.; Zheng, S.; Sun, W. Beam manipulation for quantum dot light-emitting diode with an Ag grating and a phase-gradient metasurface. *Opt. Express* **2022**, *30*, 28345–28357. [CrossRef] [PubMed]
45. Zhang, G.; Wang, C.; Bing, C.; Huang, Z.; Wang, J.; Zhang, B.; Xu, K. Polarized GaN-based LED with an integrated multi-layer subwavelength structure. *Opt. Express* **2010**, *18*, 7019–7030. [CrossRef] [PubMed]
46. Ma, M.; Meyaard, D.S.; Shan, Q.; Cho, J.; Schubert, E.F.; Kim, G.B.; Sone, C. Polarized light emission from GaInN light-emitting diodes embedded with subwavelength aluminum wire-grid polarizers. *Appl. Phys. Lett.* **2012**, *101*, 061103. [CrossRef]
47. Huang, J.P.; Xu, Y.; Zhou, B.R.; Zhan, H.M.; Cao, P.; Wang, J.X.; Li, J.M.; Yi, X.Y.; Pan, A.L.; Wang, L.C. Linearly polarized light emission from GaN micro-LEDs for 3D display. *Appl. Phys. Lett.* **2023**, *122*, 111107. [CrossRef]
48. Zhang, G.; Guo, X.; Ren, F.; Li, Y.; Liu, B.; Ye, J.; Ge, H.; Xie, Z.; Zhang, R.; Tan, H.H.; et al. High-brightness polarized green InGaN/GaN light-emitting diode structure with Al-coated p-GaN grating. *ACS Photonics* **2016**, *3*, 1912–1918. [CrossRef]
49. Zhou, L.; Zhou, Y.; Fan, B.; Nan, F.; Zhou, G.; Fan, Y.; Zhang, W.; Ou, Q. Tailored Polarization Conversion and Light-Energy Recycling for Highly Linearly Polarized White Organic Light-Emitting Diodes. *Laser Photonics Rev.* **2020**, *14*, 1900341. [CrossRef]
50. Wu, Y.; Yang, Y.; Li, T.; Huang, S.; Huang, H.; Wen, S. Stretchable and foldable waveplate based on liquid crystal polymer. *Appl. Phys. Lett.* **2020**, *117*, 263301. [CrossRef]
51. Jia, J.; Cao, X.; Ma, X.; De, J.; Yao, J.; Schumacher, S.; Liao, Q.; Fu, H. Circularly polarized electroluminescence from a single-crystal organic microcavity light-emitting diode based on photonic spin-orbit interactions. *Nat. Commun.* **2023**, *14*, 31. [CrossRef]

 *micromachines* 

Article

Photoluminescent Microbit Inscripion Inside Dielectric Crystals by Ultrashort Laser Pulses for Archival Applications

Sergey Kudryashov *, Pavel Danilov, Nikita Smirnov, Evgeny Kuzmin, Alexey Rupasov, Roman Khmelnitsky, George Krasin, Irina Mushkarina and Alexey Gorevoy

Lebedev Physical Institute, 119991 Moscow, Russia; danilovpa@lebedev.ru (P.D.); smirnovna@lebedev.ru (N.S.); kuzmine@lebedev.ru (E.K.); rupasovan@lebedev.ru (A.R.); khmelnitskyra@lebedev.ru (R.K.); krasingk@lebedev.ru (G.K.); i.mushkarina@lebedev.ru (I.M.); a.gorevoy@lebedev.ru (A.G.)
* Correspondence: kudryashovsi@lebedev.ru

Abstract: Inscription of embedded photoluminescent microbits inside micromechanically positioned bulk natural diamond, LiF and CaF_2 crystals was performed in sub-filamentation (geometrical focusing) regime by 525 nm 0.2 ps laser pulses focused by 0.65 NA micro-objective as a function of pulse energy, exposure and inter-layer separation. The resulting microbits were visualized by 3D-scanning confocal Raman/photoluminescence microscopy as conglomerates of photo-induced quasi-molecular color centers and tested regarding their spatial resolution and thermal stability via high-temperature annealing. Minimal lateral and longitudinal microbit separations, enabling their robust optical read-out through micromechanical positioning, were measured in the most promising crystalline material, LiF, as 1.5 and 13 microns, respectively, to be improved regarding information storage capacity by more elaborate focusing systems. These findings pave a way to novel optomechanical memory storage platforms, utilizing ultrashort-pulse laser inscription of photoluminescent microbits as carriers of archival memory.

Keywords: fluorides; diamond; ultrashort-pulse laser; direct laser inscription; photoluminescent microbits; vacancy clusters

Citation: Kudryashov, S.; Danilov, P.; Smirnov, N.; Kuzmin, E.; Rupasov, A.; Khmelnitsky, R.; Krasin, G.; Mushkarina, I.; Gorevoy, A. Photoluminescent Microbit Inscripion Inside Dielectric Crystals by Ultrashort Laser Pulses for Archival Applications. *Micromachines* **2023**, *14*, 1300. https://doi.org/10.3390/mi14071300

Academic Editors: Zeheng Wang and Jingkai Huang

Received: 25 May 2023
Revised: 16 June 2023
Accepted: 21 June 2023
Published: 24 June 2023

1. Introduction

Photoluminescence (PL) is one of the most important optical processes, underlying relaxation of two-level quasi-molecular systems upon their complementary optical excitation [1]. Even single PL photons could be acquired and spatially resolved much easier than differential absorption of single photons. As a result, PL characterization in 2D- or 3D-scanning confocal micro- or nano-spectroscopy mode became an enabling tool for probing local molecular or crystalline structures [2,3], or electromagnetic near-fields [4,5].

Ultrashort-pulse lasers proved to work as a versatile tool for time-resolved and/or nonlinear spectroscopy [6,7], precise surface nano- and micro-machining of any—absorbing or transparent—materials [8,9], micro-modification and inscription inside bulk transparent media [10–12]. In the latter case, (sub)microscale laser modification of molecular or crystalline structures and related PL spectra underlies facile and robust encoding of bulk diamonds for their tracing applications in identifying synthetic diamonds from natural ones in large commercial diamond collections [13], protecting trademarks of high-quality natural (potentially, synthetic too) diamond manufacturers [14], limiting commercial trading and marketing of illegal diamonds. This PL-based encoding appears unique to diamonds, where other popular encoding technologies—ablation fabrication of optically-contrasted (sub)microscale voids [15] or ablative birefringent nanogratings [16,17]—do not work in the ultra-hard diamond lattice, tending to be better for graphitization [18], while PL read-out is simpler and more sensitive. Similarly, many other crystalline scintillators and luminophores undergo ultrashort-pulse laser modification of their crystalline structures and related PL spectra [19–21], which is potentially promising for 3D optical encoding (writing/read-out) applications in storage devices (5D optical storage for specific advanced technologies [16]). Such 3D optical memory storage in bulk transparent media is an evergreen dream since the 1990s or even earlier [22], being highly promising and competitive compared to the previous 2D surface laser patterning compact disk (CD) and digital versatile (video) disk (DVD) technologies (common diameter—120 mm, thickness—1 mm, capacity—up to 17 GB for double-side, double-layer disks) utilizing 650 nm writing lasers, and the present 405 nm laser Blu-ray disk technology, supporting storage capacity up to 128 GB per four-layer disk. Other modern storage technologies enable even higher capacities—up to several TB, while possessing other technical advantages and drawbacks. Hence, rewritable, ultrahigh-capacity, low-power or autonomous high-speed memory, remaining robust to radiation, humidity and thermal shocks is still needed for quick-access and archival applications.

In this work we report a brief experimental evaluation study of natural diamond, LiF and CaF_2 crystals as optical platforms for microscale photoluminescent encoding by ultrashort-pulse lasers for micromechanically-accessed archival optical storage, of their mechanisms of laser-induced color-center inscription, tests of potential 3D memory capacity and thermal stability.

2. Materials and Methods

In these studies, a 2 mm thick colorless brick of IaA-type natural diamond (total concentration of nitrogen atoms $\approx$ 130 ppm), and 5 mm thick slabs of undoped LiF and CaF_2 crystals grown by Bridgman–Stockbarger method were utilized, being optically transparent at the writing laser wavelength of 525 nm (Figure 1a). The samples were characterized in the spectral range of 350–750 nm by room-temperature (RT) optical transmission microspectroscopy (Figure 1b), using an ultraviolet (UV)—near-IR microscope-spectrometer MFUK (LOMO, Saint-Petersburg, Russia). Inscription inside these bulk crystals at the depths of 100 μm (fluorides) and of 120 μm (diamond) in their transparency spectral regions (Figure 1b), accounting for their 525 nm refractive indexes of 2.4 (diamond), 1.4 (LiF) and 1.4 (CaF_2) [23], was performed by second-harmonic (525 nm) pulses of the TEMA Yb-crystal laser (Avesta Project, Moscow, Russia) with the pulsewidth (full width at a half maximum) of 0.2 ps, repetition rate of 80 MHz split in pulse bunches of 0.05 $\div$ 10 s duration by a mechanical shutter, and 50 nJ (average power—4 W) maximum output pulse energy E in the TEM_{00} mode. The 525 nm laser pulses with variable energies E up to 50 nJ were focused in a sub-filamentary regime (the 515 nm, 0.3 ps laser filamentation threshold energy $\approx$300 nJ (diamond) and 260 nJ (CaF_2) [24]) by a 0.65 NA micro-objective into $\approx$1 μm wide spots (1/e-intensity diameter) inside the crystals, providing the peak laser fluence <6 J/cm^2 and peak laser intensity <30 TW/cm^2. The samples were mounted on a computer-driven three-dimensional motorized micropositioning stage and exposed in separate positions with variable transverse spacings in the range of 1–5 microns and longitudinal spacings in the range of 1–28 microns.

Since different alkali or alkali-earth fluorides are rather similar during high-temperature annealing due to high vacancy mobility (activation energy for diffusion ~0.1 eV [25]) and low-temperature aggregation [26], annealing of fluoride samples was performed only for the LiF sample at different temperatures in the range of 25–300 °C (20 min temperature ramp, 30 min stationary heating), using a temperature-controlled mount for Raman microspectroscopy, while the diamond sample was annealed in an evacuated oven for 1 h at different temperatures in the range of 25–1200 °C.

In our characterization studies, top-view and side-view (cross-sectional) photoluminescence imaging at the 532 nm continuous-wave pump laser wavelength and 100× magnification (NA = 1.45, spatial resolution ~1 μm) was performed by means of a Confotec MR520 3D-scanning confocal photoluminescence/Raman microscope (SOL Instruments, Minsk, Belarus) to measure relative intensity, spatial dimensions and optical PL-acquired separation of PL microbits (Figure 2).

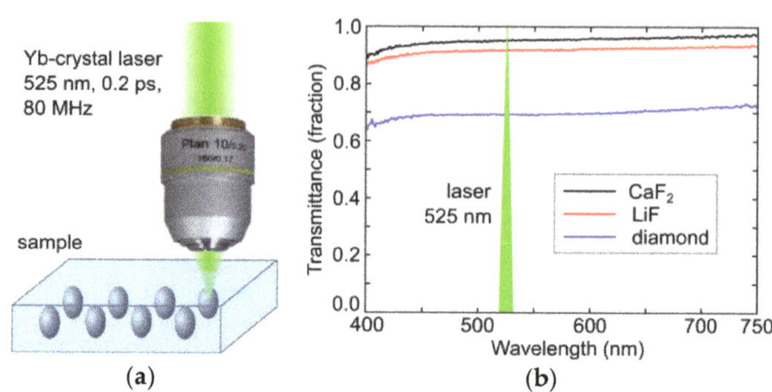

Figure 1. (**a**) Laser inscription setup, sketching the bulk PL-microbit inscription procedure; (**b**) transmittance spectra of natural diamond, LiF and CaF_2 crystals, with the laser writing wavelength of 525 nm shown in their transparency spectral region by the green triangle.

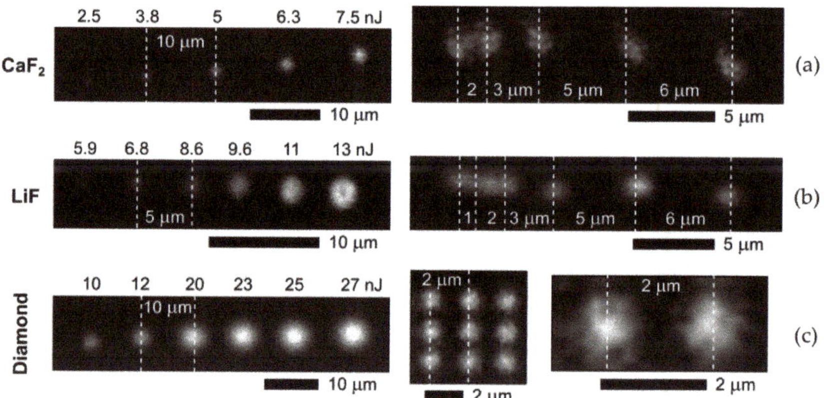

Figure 2. (**left**) Top-view PL images of linear slices of square PL microbit arrays inscribed at different laser conditions and acquired at 755 nm in CaF_2 (**a**), at 650 nm in LiF (**b**), at 650 nm in natural diamond (**c**). (**right**) Their corresponding front-view images of neighboring PL microbits inscribed in these dielectrics at different lateral separations, varying in the range of 1–6 μm.

3. Experimental Results and Discussion

3.1. Inscription of Photo-Luminescent Microbits

PL microbits were inscribed in the bulk crystalline CaF_2 and LiF slabs, as well as in the diamond plate, at different pulse energies (Figure 2a–c, left side). Specifically, the PL microbits inside the CaF_2 slab exhibit the corresponding energy-dependent microbit dimensions above the inscription threshold value of ≈3 nJ at the exposure of 10^{7-}–10^9 pulses/microbit. Similarly, the PL microbits were inscribed inside the LiF slab in the same energy range, while the threshold energy appears considerably higher (≈5 nJ, Figure 2b, left side) at the exposure of 10^{7-}–10^9 pulses/microbit, reflecting the higher bandgap energy of 13.0–14.2 eV in LiF [27,28], comparing to CaF_2 with 11.5–11.8 eV [29,30]. Finally, in the diamond plate, the PL microbits were inscribed as a function of laser pulse energy, demonstrating their increasing dimensions at the exposure of 10^{7-}–10^9 pulses/microbit (Figure 2c, left side). Surprisingly, contrary to our expectations, the inscribed microbits appear inhomogeneous at higher magnifications (Figure 2a–c, right side) because of the non-linear photoexcitation/damage character and well-known high degree of clustering—up to nanoscale—for fluorine atoms to form dislocation loops [26] or around dislocations in the fluoride crystals [25]. In diamonds, such segregation of vacancies and interstitials also occurs in the form of multi-vacancies (voids) [31] or interstitial aggregates in B2-centers [31]. In the same line, these microbits look diffuse owing to the low diffusion energies of ~0.1 eV of interstitials and vacancies [26], facilitating room-temperature internal segregation and external collateral spreading of point-defect concentrations in the PL microbits.

In terms of spatial resolution of the PL microbits during the laser inscription process, even at low above-threshold pulse energies these features could be more or less resolved only at their 2 μm separation (Figure 2a–c, right side). The main reason for such moderate lateral (transverse) resolution is apparently the initial focal 1/e-diameter of 1 μm at the focusing NA = 0.65 (see Section 2—Materials and Methods), additionally increased by ≈1 μm lateral diffusion length of fs-laser generated electron-hole plasma during its electron-lattice thermalization over 1–2 picoseconds [32]. The corresponding point beam stability upon the focusing could result in negligible lateral displacements of ~10 nm. As a result, distinct resolution of the neighboring PL microbits becomes possible for their lateral separations, exceeding 2 μm distance. Meanwhile, it could be considerably improved till ~1–1.5 μm, utilizing specially designed high-NA (0.75–0.9) air focusing micro-objectives. Below, in Section 3.4, the longitudinal (interlayer) spatial resolution will be tested in the case of a brightly luminescent LiF crystal to evaluate the potential optical storage capacity of PL microbit arrays.

3.2. Photo-Luminescence Spectra of Microbits: Atomistic Inscription and Annealing Mechanisms

Typical PL spectra acquired in the laser-inscribed microbits by 3D-scanning confocal PL micro-spectroscopy are presented in Figure 3 in comparison to the corresponding spectra of the background non-modified materials. Specifically, the CaF_2 slab exhibits the strongly enhanced PL yield in the region of 650–850 nm, peaked at 740 nm (Figure 3a). Though PL spectra of electronic excitations in fluorides are rather flexible due to high mobility of Frenkel defects and the multitude of their complexes [25,33,34], the observed peak could be assigned to some of these vacancy aggregates ($F_x^{0,+}$, where F is the fluorine vacancy with the trapped electron, x > 2 and upper indexes "0,+" denote the charged states) [25,26]. Similarly, in the LiF slab, the increased PL band in the range of 550–750 nm could be assigned to F_2 (peak at 670 nm [33]) and F_3 (peak at 650 nm [33]) centers, while the emerging PL band with its peak at 800–850 nm could also related to some as yet unknown $F_x^{0,+}$ centers [25,33,34].

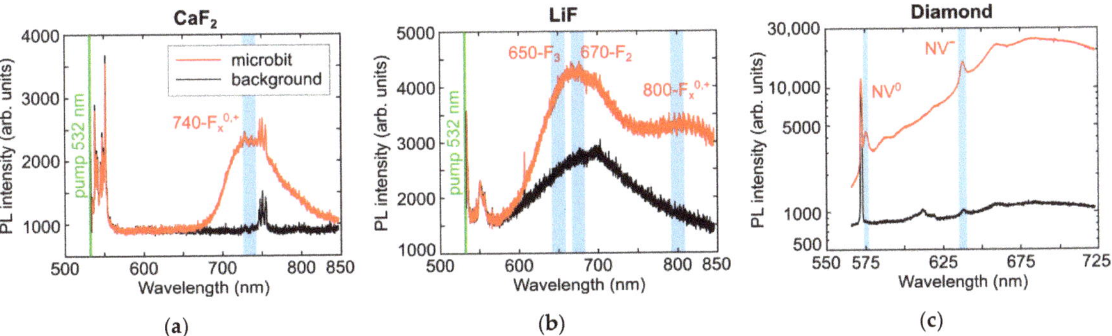

Figure 3. PL spectra of separate PL microbits (red curves) inscribed in CaF_2 (**a**), LiF (**b**) and natural diamond (**c**) regarding their background spectra of the unmodified materials (dark curves).

Finally, the observed, strongly—by one order of magnitude—enhanced PL band in the micromark inscribed inside the diamond slab exhibits the main spectral features, representing the neutral (NV^0, zero-phonon line, ZPL, at 575 nm [31]) and negatively charged nitrogen-vacancy (NV^-, zero-phonon line at 637 nm [31]) centers of substitutional nitrogen atoms with a photo-generated vacancy, as well as their red-shifted phonon replica.

Atomistic processes underlying the observed laser-induced transformations of PL spectra in LiF and CaF_2 are supposed to be associated with aggregation of mobile neutral (I-center [25,26]) and negatively charged (F-center [25,26]) vacancies, along with fluorine neutral (H-center [25,26]) and negatively charged (α-centers [25,26]) interstitials (Equation (1)), approaching to hundreds of aggregated defects usually concentrated in dislocation loops [25,26]. Likewise, in the diamond plate, the mobile photo-generated vacancies could be trapped by substitutional nitrogen atoms (C-centers [31]), resulting in well-known neutral or charged NV complexes [18,31] (Equation (1)):

$$florides: H + H \rightarrow H_2, H + V_K \rightarrow F_3^-,$$

$$diamond: N_S + V^0 \rightarrow NV. \quad (1)$$

Equation (1)—Atomistic processes, resulting in photo-induced vacancy complexes in fluorides and diamond.

In the same line, one can see strong stationary annealing of mobile vacancy-related color centers in LiF already at temperatures elevated by 200–300 °C (Figure 4a), almost deleting the microbit signal. In contrast, in denser and more rigid diamond lattice the Frenkel vacancies anchored by C-centers, remain rather stable even at high temperatures, approaching 1200 °C (Figure 4b).

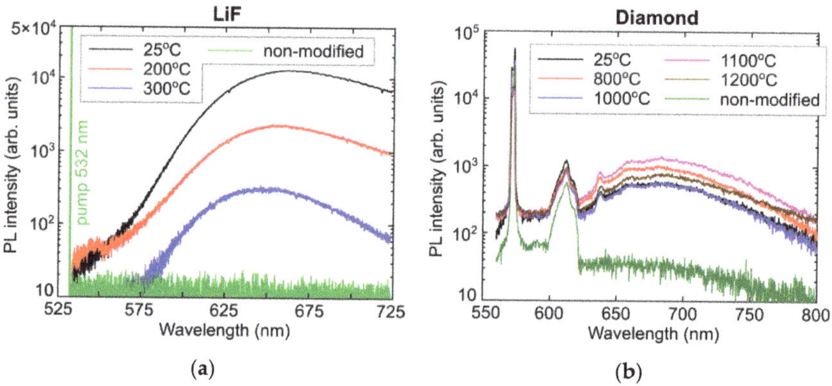

Figure 4. PL spectra of LiF (**a**) and natural diamond (**b**) upon annealing in the corresponding different temperature ranges, regarding the unannealed non-modified materials.

3.3. Photogeneration of Frenkel Pairs of Point Defects in LiF during Atomistic Inscription

PL yield at 670 nm—in the peak related to F_2-centers—was used to track fs-laser photogeneration of Frenkel pairs in LiF, underlying the formation of these centers. As can be seen in Figure 5a,c, the PL yield in LiF exhibits the non-linear (power slope in the range of ≈3.3–3.8) monotonic dependence on pulse energy E = 2.5–13 nJ (peak fluence ≈ 0.3–1.7 J/cm^2, peak intensity ≈ 1.5–9 TW/cm^2) (previously—in diamond [35]) and exposure of (4–800) × 10^6 pulses/spot (at room temperature, Figure 5b,d). Moreover, the abovementioned annealing effect at the temperatures of 200 °C and 300 °C results not only in the decreased PL intensity at 670 nm (Figure 5c), but also in the different exposure trends (Figure 5d) apparently related to cumulative heating of the material at the ultra-high 80 MHz exposure of the static sample, which is well-known to be favorable for self-trapped exciton stabilization via Frenkel pair formation [25]. The cumulative heating effect is more pronounced at room temperature (Figure 5d), while the elevated temperatures make it less distinct.

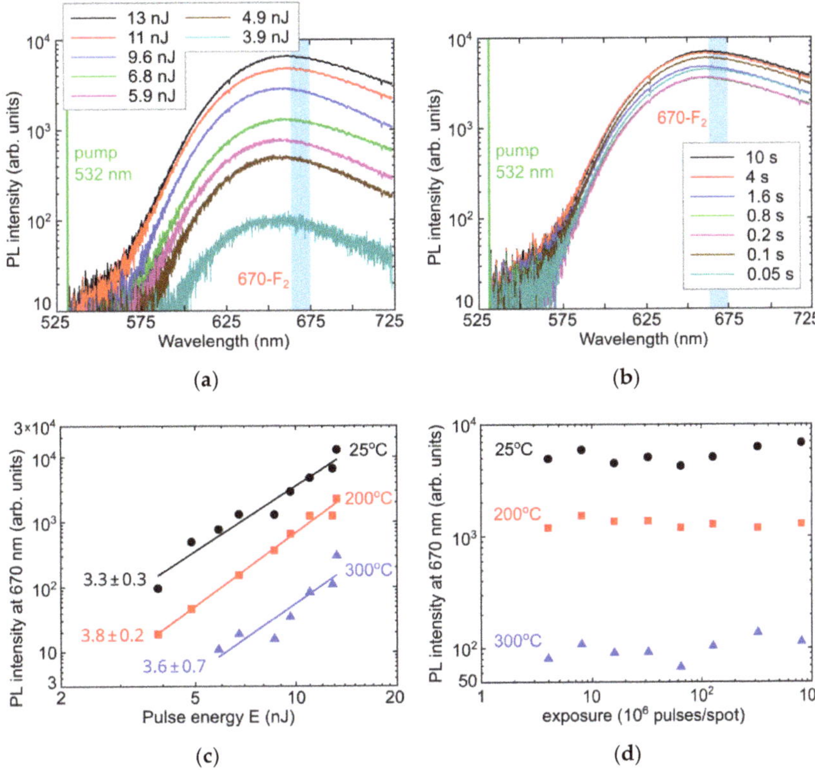

Figure 5. (**a**) PL spectra of microbits in LiF inscribed at variable pulse energy (see the frame inset) and the fixed exposure of 10 s (×80 MHz), spectral assignment after [33]; (**b**) PL spectra of microbits in LiF inscribed at the variable exposures (see the frame inset) and the fixed pulse energy of 13 nJ (spectral assignment after [33]); (**c**) PL intensity of 670 nm (F_2-center [33]) peak in the spectra as a function of pulse energy (peak fluence—0.2–2.4 J/cm^2, peak intensity—1–12 TW/cm^2) at the maximal exposure of 10 s without annealing (25 °C, black circles) and after annealing at 200 °C (red squares) and 300 °C (blue triangles) as well as their linear fitting curves of the same colors with the corresponding slopes; (**d**) PL intensity of 670 nm (F_2-center [33]) peak in the spectra as a function of exposure at the pulse energy of 13 nJ without annealing (25 °C, black circles) and after annealing at 200 °C (red squares) and 300 °C (blue triangles).

We have analyzed the observed PL yield at 670 nm vs. pulse energy E in LiF (Figure 5c), representing the concentration of F_2-centers in the probed confocal volume, alike to our previous similar studies of NV-center yield upon fs-laser exposure in diamond [35]. According to high bandgap energy of E_{dir}(Γ-point) $\approx$ 13.0–14.2 eV in LiF [27,28], formation of F_2-centers requires either $N = E_{dir}/\hbar\omega \approx 6$ photons at the 525 nm wavelength (photon energy $\hbar\omega \approx 2.4$ eV), or "hot" non-equilibrium electron of this energy (effectively, considerably higher to fulfill both the quasi-momentum and energy conservation laws). The evaluated laser-induced prompt ponderomotive enhancement of the bandgap [36,37], $U_p = e^2E^2/(4m_{opt}\omega^2)$, is minor (<1 eV) in the utilized intensity range of 9–30 TW/cm^2 (electric field strength E = 15–30 MV/cm) for the arbitrary optical mass of electron-hole pair $m_{opt} = m_e m_h/(m_e+m_h) = m_0/2$, assuming $m_e,m_h = m_0$ (free-electron mass).

Recently, for such analysis of electron-hole plasma and PL dynamics, a kinetic rate model for electron-hole plasma density ρ_{eh} was enlighteningly used in the common form [38], including (1) ultrafast, pulsewidth-limited multiphoton (cross-section σ_N), (2) impact ionization (coefficient α), (3,4) fast picosecond three-body Auger (coefficient γ) and slow binary radiative (coefficient β) recombination, as well as (5) fast self-trapping of electron-hole pairs (excitons, characteristic time τ_{str}) to produce one color center per self-trapped excitons [25] as the consecutive terms, respectively, and describing the corresponding continuous-wave-laser pumped PL yield Φ_{PL} as follows (Equation (2)) [32]:

$$\frac{d\rho_{eh}}{dt} = \sigma_N I^N + \alpha I \rho_{eh} - \gamma\rho_{eh}^3 - \beta\rho_{eh}^2 - \frac{\rho_{eh}^2}{\tau_{str}}, \Phi \propto \int \rho_{eh}^2 dt \quad (2)$$

$$\frac{d\rho_{eh}}{dt} = \sigma_6 I^6 - \gamma\rho_{eh}^3 - \frac{\rho_{eh}^2}{\tau_{str}}, \sigma_6 I^6 \approx \gamma\rho_{eh}^3, \rho_{eh} \propto I_0^2, \Phi \propto \int \rho_{eh}^2 dt \propto I_0^4. \quad (3)$$

Equations (2) and (3)—Kinetic rate equations for electron-hole plasma and related PL yield Φ_{PL} of the F_2-centers produced via exciton self-trapping [25]: (2) general form, (3) case-specific form for our experiments in LiF.

In the case of LiF, Equation (2) could be presented in the case-specific form (Equation (3)), where only six-photon ionization and Auger recombination balance each other, while excitonic self-trapping accompanies the electron-hole plasma relaxation. As a result, PL yield could follow the non-linear dependence on the peak fs-laser intensity I_0 with the power slope ≈4, being consistent with the measured values of 3.3–3.8 (Figure 5c). For the used nJ-level pulse laser energies, strong electron-hole plasma absorption is not achieved [32,39], thus enabling rather delicate inscription of PL nano- and microbits.

For comparison, in diamond, both the energy and exposure dependences of the NV^0 and NV^- color center yield exhibit non-linear (Figure 6a,c) and linear (Figure 6b,d) trends, respectively. Only linear dependence of the PL intensity on exposure time indicated the proceeding, unsaturated accumulation of the color centers. However, in terms of the pulse energy, PL intensity of NV^- centers exhibit highly-nonlinear yield (power slope—5.5 ± 0.2), while the corresponding weaker NV^0 peak (Figure 6a,b) rises similarly above the noise level at higher pulse energies (Figure 6c).

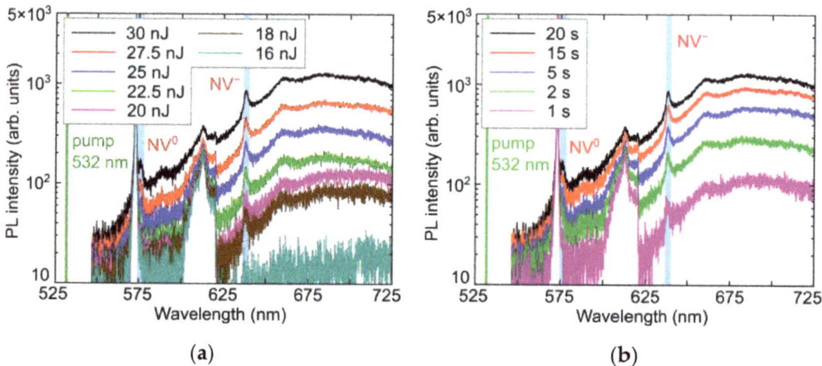

Figure 6. *Cont.*

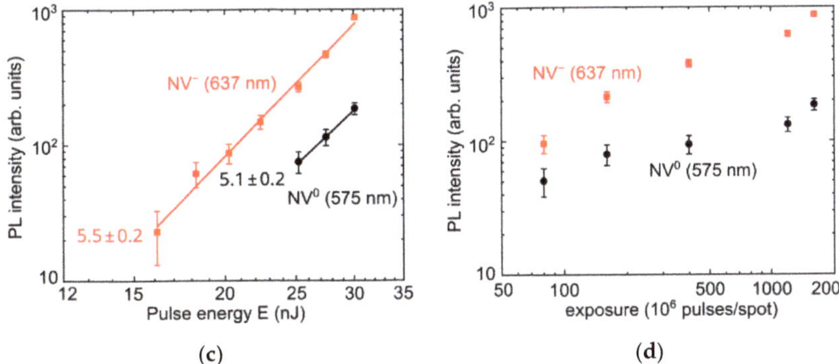

Figure 6. (**a**) PL spectra of microbits in diamond inscribed at variable pulse energy (see the frame inset) and fixed exposure of 20 s (×80 MHz), spectral assignment after [31]; (**b**) PL spectra of microbits in diamond inscribed at variable exposure (see the frame inset) and fixed pulse energy of 30 nJ (spectral assignment after [31]); (**c**) PL intensity of NV^0 (575 nm [31], black circles) and NV^- (637 nm [31], red squares) peaks in the spectra as a function of pulse energy at exposure of 20 s and linear fitting curve for NV^- with its slope indicated; (**d**) PL intensity of NV^0 (black circles) and NV^- (red squares) peaks in the spectra as a function of exposure at energy of 30 nJ.

Similarly to Equations (2) and (3), in the case of diamond, its weak non-linear photoexcitation process across the minimal direct Γ-point bandgap of 7.3 eV, which is distinct in Figure 6c, could be represented in the following familiar form [32]

$$\frac{d\rho_{eh}}{dt} = \sigma_3 I^3, \ \ \rho_{eh} \propto I_0^3, \ \ \Phi \propto \int \rho_{eh}^2 dt \propto I_0^6. \quad (4)$$

Equation (4)—Kinetic rate equations for three-photon excitation of electron-hole plasma and related PL yield Φ_{PL} produced via exciton self-trapping in diamond [31].

Here, marginal photogenerated electron-hole pairs become intermixed in the EHP, losing their initial correlation during the photogeneration and appear independently with the overall 2N-fold slope in the excitonic recombination [32], preceding NV-center formation. The observed difference in the fs-laser driven formation of Frenkel I–V pairs in LiF and diamond is apparently related to their drastically differing ionicity, favorable for excitonic self-trapping in fluorides [25,26], as compared to the predominating electron(hole)-lattice interactions in diamond [32].

3.4. Evaluation of Storage Capacity Utilizing Photo-Luminescent Microbit Arrays

Finally, we have performed experimental evaluation of PL microbit density, which is the key characteristic of archival optical storage. Above, we inscribed PL bits in the natural diamond, LIF and CaF_2 samples with ≥2 µm lateral separation, exceeding the micropositioning accuracy, which could be easily resolved in the PL images (Figure 2). Furthermore, we undertook inscription and confocal PL visualization of separate linear arrays of PL microbits with variable vertical (depth) separation changed in the range of 1–28 µm (Figure 7), in order to evaluate the minimal resolvable vertical separation. PL visualization was performed by means of Olympus (40×) and Nikon (100×) microscope objectives with vertical resolution Δz = 1 or 2 µm, respectively (Figure 7a,b). The corresponding side-view PL imaging results for the same microstructure of paired linear arrays are presented in Figure 7c,d.

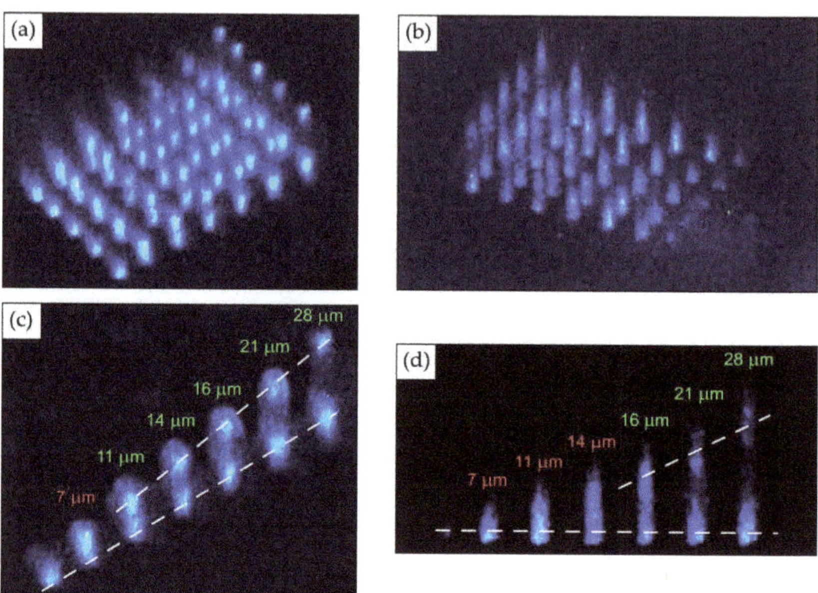

Figure 7. PL imaging of stair-like set of pairs of linear microbit arrays in LiF with variable intra-pair vertical separation in the range of 1–28 μm using Olympus (**a**,**c**) and Nikon (**b**,**d**) microscope objectives with the vertical resolution Δz = 1 or 2 μm, respectively: (**a**,**b**) 3D view; (**c**,**d**) side-view images of neighboring PL microbit arrays, showing their resolvable separation, starting from 11 μm (**c**) or 16 μm (**d**).

Here, one can find that the pairs of microbit lines become visibly separable, starting from 11 μm for the 1 μm resolution visualization (Figure 7c), while only at 16 or 21 μm—for the 2 μm resolution visualization (Figure 7d). Hence, accounting for the appropriately resolvable 2 μm intra-layer separation of microbits and their 11 μm inter-layer separation, one can evaluate the bulk microbit density of 25 Gbits per cubic centimeter for the simple cubic lattice of microbits, i.e., about 3 Tbits per disk for the 120 mm diameter of the standard CD or DVD disks and the 10 mm thickness. This optical storage capacity is comparable to previous optical memory writing technologies (visible microvoids [15], birefringent nanotrenches [16,17], photoluminescent microbits [40]), but they have clear benefits in the confocal non-linear memory read-out due to non-destructive laser inscription technology. Moreover, considerable, few-fold additional increase in the storage capacity could be achieved by higher-NA (NA > 0.65) inscription and other advanced optical means.

4. Conclusions

In this study, bulk high-NA inscription (writing) of photo-luminescent microbits in LiF, CaF_2 and diamond crystals, as a delicate laser micromachining process, was performed by means of ultrashort-pulse laser and tested (read-out) by 3D-scanning confocal photoluminescence micro-spectroscopy. Preliminarily, optical storage density for the simple cubic lattice of microbits was evaluated as high as 25 Gbits/cm^3, provided by the precise micromechanical positioning, but could be few-fold increased by using more sophisticated optical focusing tools during the encoding procedure. The underlying photo-luminescent color centers were identified in the fluorides (fluorine vacancy-based F_2-centers and similar vacancy-based specifies) and diamond (carbon vacancy-based NV-centers) by PL micro-spectroscopy, while their laser inscription mechanism was revealed in fluorides for the first time, comparing to the more or less known mechanisms for synthetic and natural diamonds. Moreover, the color centers could be easily annealed in the fluorides at moderate temperatures of 300 °C due to the high lability of the centers and room-temperature mobility of their atomistic constituents, comparing to the relatively robust color (NV) centers in diamond, persisting even at rather elevated temperatures of ≈1200 °C. Our first-step research highlights the way for potential implications of laser-inscribed photo-luminescent microbits in archival optical storage with micromechanical access for its read-out.

Author Contributions: Conceptualization, S.K., P.D. and A.G.; methodology, I.M.; validation, E.K.; investigation, A.R., P.D., N.S., R.K. and E.K.; resources, A.R., G.K. and R.K.; writing—original draft preparation, S.K.; writing—review and editing, A.G. and S.K.; visualization, P.D., E.K. and N.S.; supervision, A.G.; project administration, S.K.; and funding acquisition, S.K. All authors have read and agreed to the published version of the manuscript.

Funding: This research was funded by the Russian Science Foundation (project No. 21-79-30063); https://rscf.ru/en/project/21-79-30063/ (accessed on 4 April 2023).

Conflicts of Interest: The authors declare no conflict of interest.

References

1. Levshin, L.; Saletskii, A. *Optical Methods for Studying Molecular Systems. Molecular Spectroscopy*; MGU: Moscow, Russia, 1994.
2. Yang, B.; Chen, G.; Ghafoor, A.; Zhang, Y.; Zhang, Y.; Zhang, Y.; Luo, Y.; Yang, J.; Sandoghdar, V.; Aizpurua, J.; et al. Sub-nanometre resolution in single-molecule photoluminescence imaging. *Nat. Photonics* **2020**, *14*, 693–699.
3. Goldschmidt, J.C.; Fischer, S. Upconversion for photovoltaics–a review of materials, devices and concepts for performance enhancement. *Adv. Opt. Mater.* **2015**, *3*, 510–535.
4. Kudryashov, S.I.; Danilov, P.A.; Porfirev, A.P.; Saraeva, I.N.; Kuchmizhak, A.A.; Rudenko, A.A.; Busleev, N.I.; Umanskaya, S.F.; Zayarny, D.A.; Ionin, A.A.; et al. Symmetry-wise nanopatterning and plasmonic excitation of gold nanostructures by structured femtosecond laser pulses. *Opt. Lett.* **2019**, *44*, 1129–1132. [CrossRef]
5. Zhizhchenko, A.Y.; Tonkaev, P.; Gets, D.; Larin, A.; Zuev, D.; Starikov, S.; Pustovalov, E.V.; Zakharenko, A.M.; Kulinich, S.A.; Juodkazis, S.; et al. Light-emitting nanophotonic designs enabled by ultrafast laser processing of halide perovskites. *Small* **2020**, *16*, 2000410. [CrossRef]
6. Kudryashov, S.I.; Kandyla, M.; Roeser, C.A.; Mazur, E. Intraband and interband optical deformation potentials in femtosecond-laser-excited $\alpha-$ Te. *Phys. Rev. B* **2007**, *75*, 085207. [CrossRef]
7. Neufeld, O.; Tancogne-Dejean, N.; De Giovannini, U.; Hübener, H.; Rubio, A. Attosecond magnetization dynamics in non-magnetic materials driven by intense femtosecond lasers. *npj Comput. Mater.* **2023**, *9*, 39.
8. Joglekar, A.P.; Liu, H.H.; Meyhöfer, E.; Mourou, G.; Hunt, A.J. Optics at critical intensity: Applications to nanomorphing. *Proc. Natl. Acad. Sci. USA* **2004**, *101*, 5856–5861. [CrossRef] [PubMed]
9. Syubaev, S.; Zhizhchenko, A.; Kuchmizhak, A.; Porfirev, A.; Pustovalov, E.; Vitrik, O.; Kulchin, Y.; Khonina, S.; Kudryashov, S. Direct laser printing of chiral plasmonic nanojets by vortex beams. *Opt. Express* **2017**, *25*, 10214–10223. [CrossRef]
10. Canning, J.; Lancry, M.; Cook, K.; Weickman, A.; Brisset, F.; Poumellec, B. Anatomy of a femtosecond laser processed silica waveguide. *Opt. Mater. Express* **2011**, *1*, 998–1008.
11. Juodkazis, S.; Nishimura, K.; Tanaka, S.; Misawa, H.; Gamaly, E.G.; Luther-Davies, B.; Hallo, L.; Nicolai, P.; Tikhonchuk, V.T. Laser-induced microexplosion confined in the bulk of a sapphire crystal: Evidence of multimegabar pressures. *Phys. Rev. Lett.* **2006**, *96*, 166101. [CrossRef]
12. Zhang, B.; Liu, X.; Qiu, J. Single femtosecond laser beam induced nanogratings in transparent media—Mechanisms and applications. *J. Mater.* **2019**, *5*, 1–14.
13. Tsai, T.H. Imaging-assisted Raman and photoluminescence spectroscopy for diamond jewelry identification and evaluation. *Appl. Opt.* **2023**, *62*, 2587–2594. [PubMed]
14. ALROSA. Identification of Diamonds Using Laser Nanomarks. Available online: https://youtu.be/X3Z_jcWowks (accessed on 11 December 2022).
15. Hong, M.H.; Luk'yanchuk, B.; Huang, S.M.; Ong, T.S.; Van, L.H.; Chong, T.C. Femtosecond laser application for high capacity optical data storage. *Appl. Phys. A* **2004**, *79*, 791–794. [CrossRef]
16. Wang, H.; Lei, Y.; Wang, L.; Sakakura, M.; Yu, Y.; Shayeganrad, G.; Kazansky, P.G. 100-Layer Error-Free 5D Optical Data Storage by Ultrafast Laser Nanostructuring in Glass. *Laser Photonics Rev.* **2022**, *16*, 2100563. [CrossRef]
17. Lipatiev, A.S.; Fedotov, S.S.; Lotarev, S.V.; Lipateva, T.O.; Shakhgildyan, G.Y.; Sigaev, V.N. Single-Pulse Laser-Induced Ag Nanoclustering in Silver-Doped Glass for High-Density 3D-Rewritable Optical Data Storage. *ACS Appl. Nano Mater.* **2022**, *5*, 6750–6756.
18. Ionin, A.A.; Kudryashov, S.I.; Mikhin, K.E.; Seleznev, L.V.; Sinitsyn, D.V. Bulk femtosecond laser marking of natural diamonds. *Laser Phys.* **2010**, *20*, 1778–1782. [CrossRef]
19. Chen, Y.-C.; Salter, P.S.; Knauer, S.; Weng, L.; Frangeskou, A.C.; Stephen, C.J.; Ishmael, S.N.; Dolan, P.R.; Johnson, S.; Green, B.L.; et al. Laser writing of coherent colour centres in diamond. *Nat. Photonics* **2017**, *11*, 77–80.

20. Chekalin, S.V.; Kompanets, V.O.; Dormidonov, A.E.; Kandidov, V.P. Influence of induced colour centres on the frequency–angular spectrum of a light bullet of mid-IR radiation in lithium fluoride. *Quantum Electron.* **2017**, *47*, 259. [CrossRef]
21. Castelletto, S.; Maksimovic, J.; Katkus, T.; Ohshima, T.; Johnson, B.C.; Juodkazis, S. Color centers enabled by direct femtosecond laser writing in wide bandgap semiconductors. *Nanomaterials* **2020**, *11*, 72.
22. Akimov, D.A.; Zheltikov, A.M.; Koroteev, N.I.; Magnitskiy, S.A.; Naumov, A.N.; Sidorov-Biryukov, D.A.; Sokolyuk, N.T.; Fedotov, A.B. Data reading with the aid of one-photon and two-photon luminescence in three-dimensional optical memory devices based on photochromic materials. *Quantum Electron.* **1998**, *28*, 547–554. [CrossRef]
23. Palik, E.D. *Handbook of Optical Constants of Solids*; Academic Press: Orlando, FL, USA, 1998.
24. Kudryashov, S.I.; Danilov, P.A.; Kuzmin, E.V.; Gulina, Y.S.; Rupasov, A.E.; Krasin, G.K.; Zubarev, I.G.; Levchenko, A.O.; Kovalev, M.S.; Pakholchuk, P.P.; et al. Pulse-width-dependent critical power for self-focusing of ultrashort laser pulses in bulk dielectrics. *Opt. Lett.* **2022**, *47*, 3487–3490. [CrossRef]
25. Song, K.; Williams, R.T. *Self-Trapped Excitons*; Springer: Berlin/Heidelberg, Germany, 1993.
26. Hayes, W.; Stoneham, A.M. *Defects and Defect Processes in Nonmetallic Solids*; Wiley: Dover, NY, USA, 2004.
27. Samarin, S.; Artamonov, O.M.; Suvorova, A.A.; Sergeant, A.D.; Williams, J.F. Measurements of insulator band parameters using combination of single-electron and two-electron spectroscopy. *Solid State Commun.* **2004**, *129*, 389–393.
28. Karsai, F.; Tiwald, P.; Laskowski, R.; Tran, F.; Koller, D.; Gräfe, S.; Blaha, P. F center in lithium fluoride revisited: Comparison of solid-state physics and quantum-chemistry approaches. *Phys. Rev. B* **2014**, *89*, 125429.
29. Ma, Y.; Rohlfing, M. Quasiparticle band structure and optical spectrum of CaF2. *Phys. Rev. B* **2007**, *75*, 205114. [CrossRef]
30. Chen, J.; Zhang, Z.; Guo, Y.; Robertson, J. Electronic properties of CaF2 bulk and interfaces. *J. Appl. Phys.* **2022**, *131*, 215302.
31. Zaitsev, A.M. *Optical Properties of Diamond: A Data Handbook*; Springer: Berlin/Heidelberg, Germany, 2001.
32. Kudryashov, S.; Danilov, P.; Smirnov, N.; Krasin, G.; Khmelnitskii, R.; Kovalchuk, O.; Kriulina, G.; Martovitskiy, V.; Lednev, V.; Sdvizhenskii, P.; et al. "Stealth Scripts": Ultrashort Pulse Laser Luminescent Microscale Encoding of Bulk Diamonds via Ultrafast Multi-Scale Atomistic Structural Transformations. *Nanomaterials* **2023**, *13*, 192.
33. Lakshmanan, A.R.; Madhusoodanan, U.; Natarajan, A.; Panigrahi, B.S. Photoluminescence of F-aggregate centers in thermal neutron irradiated LiF TLD-100 single crystals. *Phys. Status Solidi (a)* **1996**, *153*, 265–273. [CrossRef]
34. Davidson, A.T.; Kozakiewicz, A.G.; Comins, J.D. Photoluminescence and the thermal stability of color centers in γ-irradiated LiF and LiF (Mg). *J. Appl. Phys.* **1997**, *82*, 3722–3729. [CrossRef]
35. Kudryashov, S.I.; Danilov, P.A.; Smirnov, N.A.; Stsepuro, N.G.; Rupasov, A.E.; Khmelnitskii, R.A.; Oleynichuk, E.A.; Kuzmin, E.V.; Levchenko, A.O.; Gulina, Y.S.; et al. Signatures of ultrafast electronic and atomistic dynamics in bulk photoluminescence of CVD and natural diamonds excited by ultrashort laser pulses of variable pulsewidth. *Appl. Surf. Sci.* **2021**, *575*, 151736.
36. Keldysh, L.V. Ionization in the field of a strong electromagnetic wave. *Sov. Phys. JETP* **1965**, *20*, 1307–1314.
37. Kudryashov, S.I. Microscopic model of electronic Kerr effect in strong electric fields of intense femtosecond laser pulses. In Proceedings of the Conference on Quantum Electronics and Laser Science, Baltimore, MA, USA, 22–27 May 2005; p. JThE26.
38. Bulgakova, N.M.; Stoian, R.; Rosenfeld, A.; Hertel, I.V.; Campbell, E.E.B. Electronic transport and consequences for material removal in ultrafast pulsed laser ablation of materials. *Phys. Rev. B* **2004**, *69*, 054102.
39. Khorasani, M.; Ghasemi, A.; Leary, M.; Sharabian, E.; Cordova, L.; Gibson, I.; Downing, D.; Bateman, S.; Brandt, M.; Rolfe, B. The effect of absorption ratio on meltpool features in laser-based powder bed fusion of IN718. *Opt. Laser Technol.* **2022**, *153*, 108263.
40. Ren, Y.; Li, Y.; Guo, K.; Cui, Z.; Wang, C.; Tan, Y.; Liu, H.; Cai, Y. Dual-modulation of micro-photoluminescence in rare-earth-doped crystals by femtosecond laser irradiation for 5D optical data storage. *Opt. Lasers Eng.* **2023**, *167*, 107612.

 *micromachines*

Article

A FIN-LDMOS with Bulk Electron Accumulation Effect

Weizhong Chen [1,2], Zubing Duan [2,*], Hongsheng Zhang [1], Zhengsheng Han [2] and Zeheng Wang [3,*]

1 College of Electronics Engineering, Chongqing University of Posts and Telecommunications, Chongqing 400065, China; chenwz@cqput.edu.cn (W.C.); zhanghs@cqupt.edu.cn (H.Z.)
2 Institute of Microelectronics, Chinese Academy of Sciences, Beijing 100029, China
3 CSIRO Manufacturing, 36 Bradfield Road, P.O. Box 218, Lindfield, NSW 2070, Australia
* Correspondence: zubingduan@foxmail.com (Z.D.); zenwang@outlook.com (Z.W.)

Abstract: A thin Silicon-On-Insulator (SOI) LDMOS with ultralow Specific On-Resistance ($R_{on,sp}$) is proposed, and the physical mechanism is investigated by Sentaurus. It features a FIN gate and an extended superjunction trench gate to obtain a Bulk Electron Accumulation (BEA) effect. The BEA consists of two p-regions and two integrated back-to-back diodes, then the gate potential V_{GS} is extended through the whole p-region. Additionally, the gate oxide W_{oxide} is inserted between the extended superjunction trench gate and N-drift. In the on-state, the 3D electron channel is produced at the P-well by the FIN gate, and the high-density electron accumulation layer formed in the drift region surface provides an extremely low-resistance current path, which dramatically decreases the $R_{on,sp}$ and eases the dependence of $R_{on,sp}$ on the drift doping concentration (N_{drift}). In the off-state, the two p-regions and N-drift deplete from each other through the gate oxide W_{oxide} like the conventional SJ. Meanwhile, the Extended Drain (ED) increases the interface charge and reduces the $R_{on,sp}$. The 3D simulation results show that the BV and $R_{on,sp}$ are 314 V and 1.84 mΩ·cm^{-2}, respectively. Consequently, the FOM is high, reaching up to 53.49 MW/cm^2, which breaks through the silicon limit of the RESURF.

Keywords: bulk electron accumulation (BEA); extended superjunction trench gate; extended drain (ED); BV and $R_{on,sp}$

Citation: Chen, W.; Duan, Z.; Zhang, H.; Han, Z.; Wang, Z. A FIN-LDMOS with Bulk Electron Accumulation Effect. *Micromachines* **2023**, *14*, 1225. https://doi.org/10.3390/mi14061225

Academic Editor: Ha Duong Ngo

Received: 2 May 2023
Revised: 24 May 2023
Accepted: 8 June 2023
Published: 10 June 2023

1. Introduction

The Lateral Double-diffused Metal–Oxide Semiconductor (LDMOS) is a very important device in power-integrated circuits and electronic power systems [1–3], which are used in many places in our daily life. The Breakdown Voltage (BV) and the Specific On-Resistance ($R_{on,sp}$) are significant parameters to evaluate the quality of devices [4–7]. For the conventional LDMOS, there is an unavoidable trade-off relationship between the BV and $R_{on,sp}$, which can be written as $R_{on} \propto BV^{2.5}$, while what we need are high BV and low $R_{on,sp}$. Baliga's Figure Of Merit (FOM) is calculated by $BV^2/R_{on,sp}$ to evaluate the device, where a higher value is better [8–10]. Many advanced theories and structures have been investigated to increase the FOM of the power devices [11–14]. For example, for the BFG LDMOS proposed in [11], the author made half of the device into a grid, and for the HKGF LDMOS proposed in [14], the authors distinguished the device drift into three parts, each surrounded by a three-dimensional High-K dielectric, both of which greatly enhanced the control ability of the device and greatly reduced the $R_{on,sp}$ of the device. The Enhanced Dielectric layer Field (ENDIF) theory can introduce a higher electric field between the top layer of silicon and the buried oxygen layer, which can obviously enhance BV [15–17]. However, it is possible to increase $R_{on,sp}$ with the application of ENDIF theory. For example, the T-SJ LDMOS proposed in [17] enhances the BV of the device by making the top layer of silicon near the drain extremely thin, but greatly reduces the volume of the device drift region, thus reducing the $R_{on,sp}$ of the device. The SuperJunction (SJ) structure can effectively increase the drift doping concentration by using N-type semiconductors and P-type semiconductors to assist each other; thus, the $R_{on,sp}$ is reduced and the BV is guaranteed simultaneously, and thus the *FOM* can be effectively improved [18–21]. However, the superjunction structure is often placed on the P-type substrate in lateral devices, and the surface superjunction region is affected by the Substrate-Assisted Depletion (SAD) effect, which results in reduced pressure tolerance. Moreover, the power FINFET with a 3D electron channel and the LDMOS with an Accumulation Extended Gate (AEG LDMOS) are also used to improve the device [22].

In this paper, the FIN-LDMOS with Bulk Electron Accumulation (BEA-LDMOS) is first proposed. The device adopts an SOI structure, which can effectively suppress the SAD effect. The bulk electron accumulation effect produced by the extended superjunction trench gate in the N-drift and the ENDIF effect produced by the Extended Drain are produced. Moreover, the assisted depletion effect is performed by the extended superjunction trench gate, which dramatically reduces the $R_{on,sp}$ while BV is guaranteed. The devices are performed by Synopsys Sentaurus, and the main physics models are as follows: Effectice Intrinsic Density, Mobility (High Field Saturation Enormal PhuMob DopingDependence), and Recombination (SRH Auger Avalanche) [23].

2. Device Structure and Mechanism

2.1. Device Structure of the BEA

The FIN-LDMOS, AEG-LDMOS, and the proposed BEA-LDMOS are compared together in Figure 1. The FIN-LDMOS greatly increases the channel area of the device by turning the one-dimensional gate into a three-dimensional one, to reduce the $R_{on,sp}$. The AEG-LDMOS structure is characterized by extending the gate, which is only near the source region in the conventional LDMOS, to the drain, separated by SiO_2. In order to extend the drain potential better, two back-to-back PN junctions are used in the extension gate to generate a charge accumulation effect at the top of the N-drift area, forming a low-resistance channel and greatly reducing the $R_{on,sp}$ of the device. However, the BV will be affected by the PN junction in the extension grid, which reduces the BV of the device. The 3-D structure of the BEA-LDMOS is shown in Figure 1. The extended superjunction trench gate is formed as follows: the two back-to-back diodes are introduced at the collector, and the P-region and N-drift are separated by the SiO_2. Moreover, the two ends of the P-region are, respectively, shortly connected to the gate electrode and the drain electrode, as shown in Figure 1a. When the device is in the on-state, the positive V_{GS} is extended through most of the P-region to the positively biased D_1, and positive V_{DS} is extended through the P-region to the positively biased D_2. Thus, the P-region is covered by the V_{GS} and V_{DS}; then, a Bulk Electron Accumulation (BEA) is generated in the drift region, as shown in Figure 1b, and the metal–insulator–semiconductor structure is composed of P-region/oxide/N-drift. It is equivalent to a low-resistance 3-D channel in the drift region. Therefore, the $R_{on,sp}$ of the device is greatly reduced. Figure 1c indicates the breakdown mechanism of the BEA-LDMOS in the off-state, and it is similar to the conventional superjunction. The P-region and the N-drift region are separated by SiO_2, and there is still a built-in electric field from the N-drift region toward the P-region, resulting in the formation of a depletion region near SiO_2. Therefore, the electric field of the N-drift is modulated, thus helping to increase BV. At the same time, the Extended Drain structure is also introduced in the drift area, which can not only introduce the high electric field near the drain region into the more voltage-resistant silicon dioxide buried layer to increase the BV of the device, but also play the role of low-resistance channel, reducing the $R_{on,sp}$ of the device. The key parameters of the devices are listed in Table 1, with a drift length L_D of 21.0 μm, depth T_D of 5.0 μm, Extended Drain length L of 9 μm, and thickness of 0.4 μm. The optimized drift doping N_{drift} of 2.2×10^{15} cm^{-3} is designed for the BEA-LDMOS.

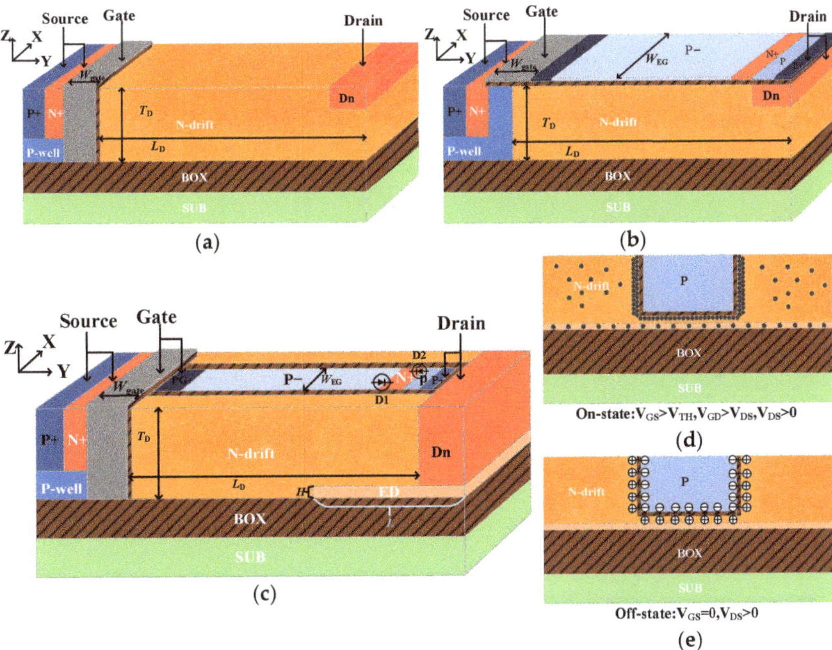

Figure 1. The three-dimensional (3D) schematic and mechanism of the three proposed devices. (**a**) FIN-LDMOS, (**b**) AEG-LDMOS, (**c**) BEA-LDMOS, (**d**) the Bulk Electron Accumulation (BEA) effect induced at the N-drift, (**e**) the assisted depletion between the P-region and N-drift in the Off-state. Diode D1 is formed by the (P−/N+) and D2 is formed by the (P/N+) junction.

Table 1. Key parameters used in simulation.

Symbol	Description	FIN-LDMOS	CON-LDMOS	AEG-LDMOS	BEA-LDMOS
L_D	drift length (μm)	21	21	21	21
T_D	drift depth (μm)	5	5	5	5
W_{gate}	Gate wideth (μm)	2	2	2	2
N_{drift}	N-drift doping (cm^{-3})	2.5×10^{15}	2.5×10^{15}	2.5×10^{15}	2.5×10^{15}
$h\text{-}top$	AEG structure thickness (μm)	--	--	0.2	--
W_{oxide}	Gate oxide width (μm)	0.1	0.1	0.1	0.1
W_{EG}	Extended gate width (μm)	--	--	2.2	0.6
L_{FIN}	FIN-Gate length (μm)	1	--	--	1
L	Extended drain lenth (μm)	--	--	--	9
H	Extended drain thickness (μm)	--	--	--	0.4

The simplified production process of the BEA-LDMOS is given as follows: The emphasis is the extended superjunction trench gate and Extended Drain. The SOI wafer in Figure 2a is injected with oxygen ions and annealed to form a silicon dioxide isolation layer. The P-well and Extended Drain are obtained by implantation, as shown in Figure 2b. The following undergo thermal oxidation to grow the isolation layer and undergo secondary ion implantation doping, as shown in Figure 2c. The subsequent processes such as metallization and passivation are compatible with conventional LDMOSs in Figure 2d.

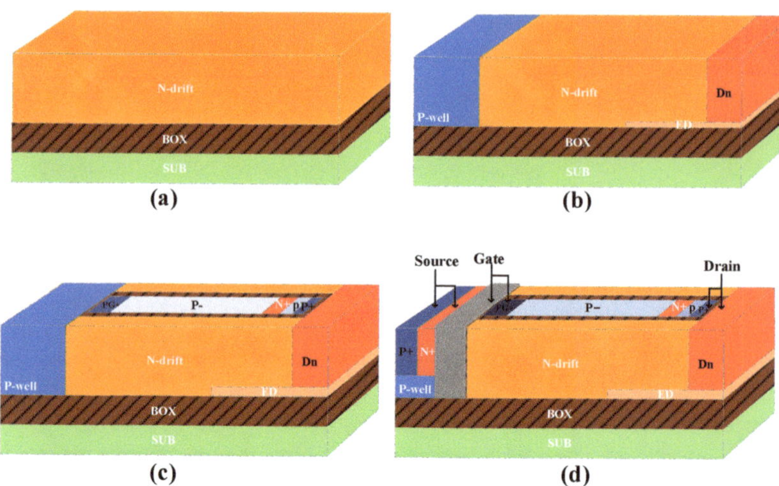

Figure 2. Simplified key process of the proposed BEA-LDMOS. (**a**) SOI substate, (**b**) Ion doping, (**c**) Thermal oxidation, (**d**) Electrode deposition.

2.2. Mainly Applied Physical Models

The main physical models used in this simulation include the carrier mobility model, carrier recombination model, and avalanche breakdown generation model [24–26]. Carrier mobility is expressed as follows:

$$\frac{1}{\mu} = \frac{\exp\left(-\frac{x}{l_{\text{cirt}}}\right)}{\mu_{\text{ac}}} + \frac{\exp\left(-\frac{x}{l_{\text{cirt}}}\right)}{\mu_{\text{sr}}} + \frac{1}{\mu_b} + \frac{1}{\mu_F} \tag{1}$$

where μ_{ac} represents the surface phonon scattering model, μ_{sr} represents the surface roughness scattering model, x represents the distance between the insulator and the semiconductor interface, μ_b represents the low-field-mobility model, and μ_F represents the high-field-mobility model.

In the simulation, we use the SRH carrier recombination model, which can accurately simulate the recombination mechanism of carrier under quantum effects. The carrier recombination model is expressed as follows:

$$R_{net}^{\text{SRH}} = \frac{np - \gamma_n \gamma_p n_{i,eff}}{T_p\left(n + \gamma_n n_{i,eff}\right) + T_n\left(p + \gamma_p n_{i,eff}\right)} \tag{2a}$$

$$\gamma_n = \frac{n}{N_C} exp(-\frac{E_{Fn} - E_C}{kT_n}) \tag{2b}$$

$$\gamma_p = \frac{p}{N_V} exp(-\frac{E_V - E_{Fp}}{kTp}) \tag{2c}$$

In the formula, T_n represents the lifetime of non-equilibrium minority electron, T_p represents the lifetime of non-equilibrium minority hole, N_C represents the effective state density of the conduction band, N_V represents the effective state density of the valence band, E_{Fn} represents the quasi-Fermi level of the conduction band, E_{Fp} represents the quasi-Fermi level of the valence band.

In order to accurately simulate the breakdown voltage of the device, the avalanche breakdown generation model is introduced in the simulation process. When the device is working in the blocking voltage, as the drain voltage continues to increase, the internal electric field of the device becomes stronger. When the maximum electric field inside the device is greater than or equal to the critical breakdown electric field of silicon, the charge multiplication effect will occur, and the leakage of the device will increase sharply, resulting in electrical breakdown of the device. The avalanche breakdown generation model is expressed as follows:

$$G^{Avalanche} = \alpha_n n v_n + \alpha_p p v_p \tag{3a}$$

$$\alpha = \gamma a e^{-\frac{\gamma b}{F}} \tag{3b}$$

$$\gamma = \frac{tanh(\frac{h\omega_{op}}{2kT_0})}{tanh(\frac{h\omega_{op}}{2kT})} \tag{3c}$$

In the formula, α is the ionization factor, it's the inverse of the mean free path, F represents the Vector mechanics, $h\omega_{op}$ represents the optical phonon energy, y represents the Dependence coefficient of phonon.

3. Results and Discussion

3.1. Control Mechanism and Bulk Electron Accumulation Effect of the BEA

Figure 3 shows the transfer, transconductance (g_{m}), and gate potential characteristics for the devices. Figure 3a shows that when V_{GS} is increased from 6 V to 16 V, and the V_{DS} of the drain is grounded, V_{GS} is extended through the whole P-region, and it is shunted by the negatively biased D_2. Figure 3b compares the transfer and g_{m} characteristics for the CON-LDMOS, AEG-LDMOS, FIN-LDMOS, and BEA-LDMOS. The peak g_{m} for the devices is 2.26, 3.16, 4.21, and 18.33 mS/mm, respectively. Because the higher peak g_{m} can be achieved by the 3-D bulk electron channel, the BEA shows the best control capability of I_{DS}.

Figure 4 shows the electron current densities of devices in the on-state. It can be seen that the electron current density of the AEG-LDMOS and BEA-LDMOS are much higher than those of the other two devices due to the existence of a charge accumulation effect. Moreover, because of the 3-D charge accumulation effect of the BEA-LDMOS, while the charge accumulation effect of the AEG is one-dimensional, the area of the low-resistance channel formed by the BEA-LDMOS is much higher than that of AEG-LDMOS, so the electron current density of the BEA-LDMOS is the largest of all four devices.

Figure 5 shows the output I_{DS}-V_{DS} characteristics of the four devices at the forward conduction. For the proposed BEA-LDMOS, the P-region is covered by V_{GS} and V_{DS} to obtain a 3-D low-resistance channel, so the linear current and saturation current in the drift region are much higher than those of the CON-LDMOS, AEG-LDMOS, and FIN-LDMOS under the same V_{DS}. Thus, stronger conductivity and ultra-low $R_{\mathrm{on,sp}}$ are achieved.

3.2. Specifics of Resistance $R_{on,sp}$ and Breakdown Voltage BV

Figure 6 shows the influence of the thickness of the top layer of silicon on $R_{\mathrm{on,\,sp}}$ and BV of the device. It can be seen from the figure that $R_{\mathrm{on,sp}}$ gradually decreases with the increase in T_{D} under different gate voltages. This can be explained by the formula for volume resistance:

$$R = R_s \frac{L}{W} \tag{4a}$$

$$R_s = \frac{\rho}{T_{\mathrm{D}}} \tag{4b}$$

where L and W are the length and width of the device channel, respectively; R_s is the resistance of the block; ρ is the resistivity; and T_{D} is the thickness of the silicon film. However, the BV first increases and then decreases with the increase in T_{D}, and the optimum BV is 314 V when T_D is 5.0 μm. This is because when T_D is small, the longitudinal breakdown voltage of the device is very low, and the BV of the device mainly depends on the longitudinal breakdown voltage. When T_D is too large, the relationship between N_{drift} and T_D does not conform to the RESURF theory [27], the device will breakdown in advance, as shown in Figure 6b, and the electric field will be 0 at half of the drift area of the device.

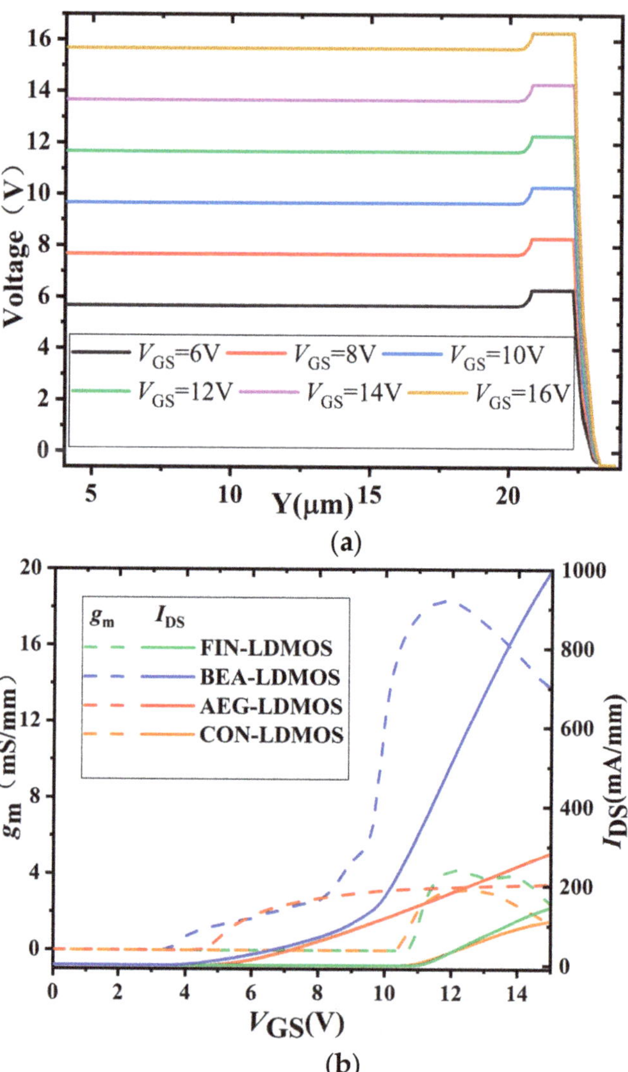

Figure 3. The gate potential V_{GS} in the on-state, transfer, and g_m characteristics for the devices. (**a**) Potential distribution along the p–n–p for the BEA-LDMOS. (**b**) Transfer and g_m of the CON-LDMOS, FIN-LDMOS, AEG-LDMOS, and BEA-LDMOS.

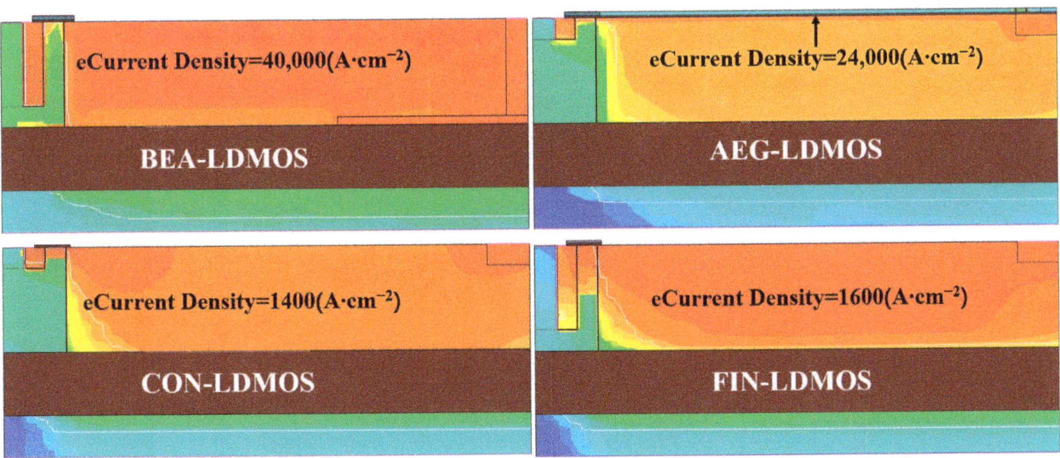

Figure 4. The electgron Current density distribution for the four devices at the On-state.

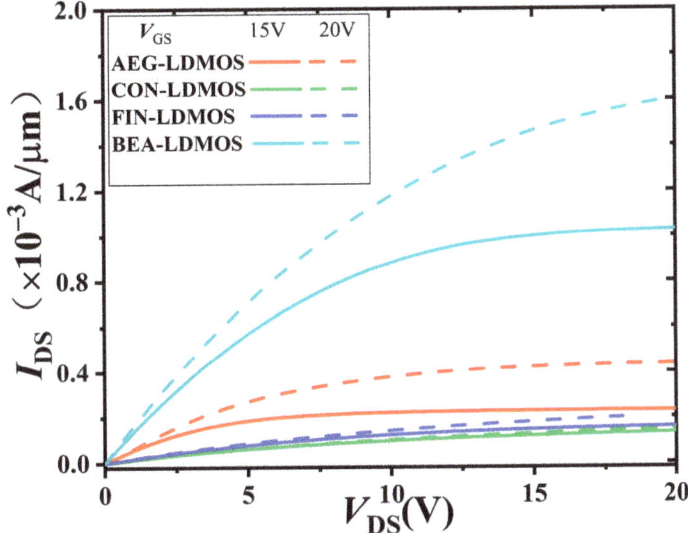

Figure 5. Output characteristics of the devices, and V_{GS} of 15 and 20 V are applied.

Figure 7 shows the influences of the doping N_{drift} on the $R_{on,sp}$ and BV for the devices. For the CON-LDMOS, the optimized BV and $R_{on,sp}$ are 329 V and 14.54 mΩ·cm^2 when the N_{drift} is 2.5 × 10^{15} cm^{-3}, respectively. For the FIN-LDMOS, the optimized BV and $R_{on,sp}$ are 323 V and 11.78 mΩ·cm^2 when the N_{drift} is 2.5 × 10^{15} cm^{-3}, respectively. For the BEA-LDMOS, the optimized BV and $R_{on,sp}$ are 314 V and 1.84 mΩ·cm^2 when the N_{drift} is 2.2 × 10^{15} cm^{-3}, respectively. It can be seen that the BV of the four devices increases first and then decreases with the increase in N_{drift}. This is because when the doping concentration is very low, the maximum electric field of the device is near the drain region, and the breakdown voltage of the device mainly depends on the heterojunction formation of the drain region and N-drift region. When the doping concentration in the N-drift region gradually increases, the PN junction formed in the drift region and the P-well also begins to participate in the voltage resistance. When the N-drift region doping concentration continues to increase, the maximum electric field of the device will appear near the source region, and the doping concentration difference between the drain and the drift region is very small. The breakdown voltage of the device mainly depends on the PN junction formed by the drift region and P-well. The breakdown voltage reaches its maximum when the two electric field spikes are almost high. The $R_{on,sp}$ of CON-LDMOS and FIN-LDMOS decreases obviously with the increase in N_{drift}, but the $R_{on,sp}$ of AEG-LDMOS and BEA-LDMOS almost does not change with the change in N_{drift}. This is because these two devices have low-resistance channels formed by the electron accumulation effect, so $R_{on,sp}$ does not depend on N_{drift}. Consequently, Baliga's Figures OF Merit (FOMs) of the CON-LDMOS, FIN-LDMOS, AEG-LDMOS, and BEA-LDMOS are calculated as 7.42 MW/cm^2, 8.86 MW/cm^2, 20.02 MW/cm^2, and 53.43 MW/cm^2, respectively.

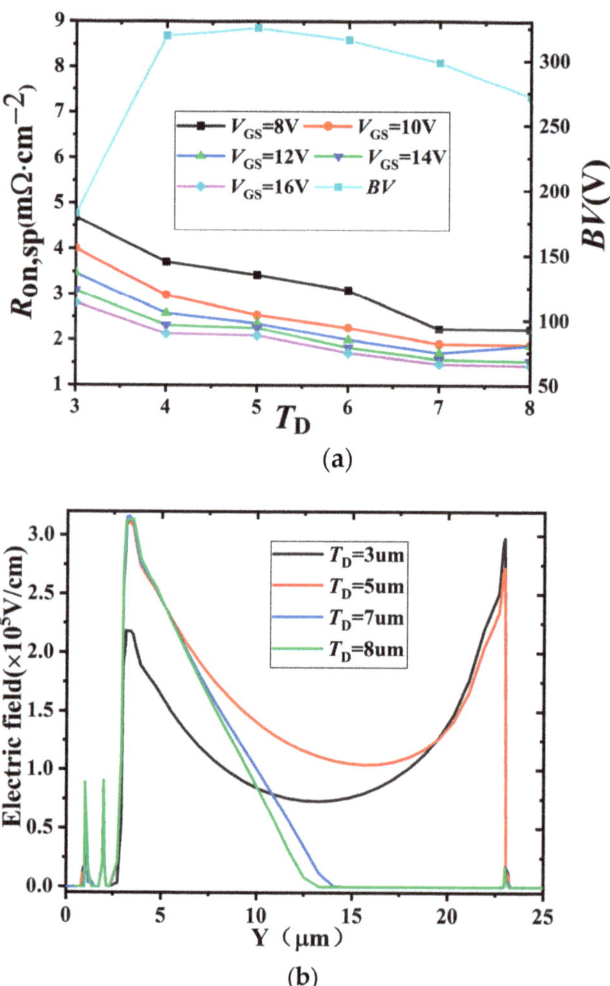

Figure 6. Influence of key parameter T_D on the $R_{on,sp}$, BV, and electric field for the BEA-LDMOS (T_D is the thickness of the N-drift). (**a**) Influence on the $R_{on,sp}$ and BV, (**b**) influence on the electric field of top layer.

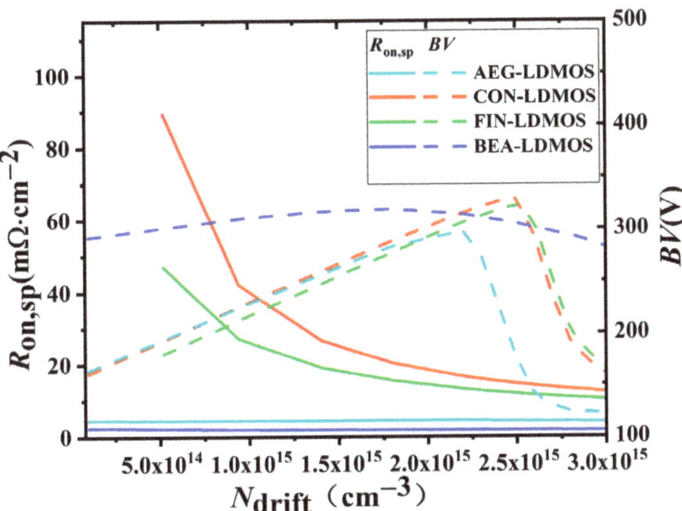

Figure 7. Effect of doping concentration of N-drift region on BV and $R_{on,sp}$ for the four devices.

Figure 8 shows the corresponding equipotential distribution at the avalanche breakdown for the four devices. The yellow area in the figure is the N-drift area, the brown area is the buried oxygen layer, the green area is the P-type substrate, the top left is the drain, the right is the source and the gate. It is noted that the BVs of the CON-LDMOS, FIN-LDMOS, AEG-LDMOS, and BEA-LDMOS are 315, 306, 297, and 314 V, respectively. The CON-LDMOS, FIN-LDMOS, and AEG-LDMOS have a similar distribution of potential lines, but there is no distribution of potential lines in the area near the drain of the BEA-LDMOS. This is because the Extended Drain introduces the higher electric field into the silicon dioxide layer, so the silicon dioxide layer below the Extended Drain has a denser distribution of the potential line. Because SiO_2 has a smaller interfacial defect density and a larger dielectric constant than silicon, SiO_2 can withstand a higher voltage and can improve the BV of the device.

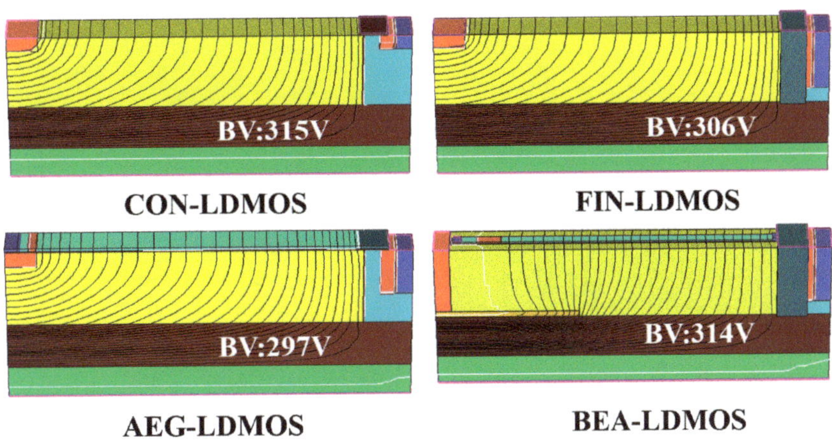

Figure 8. The 3-D equipotential contours at the avalanche breakdown at the same N_{drift}.

Figure 9 demonstrates the electric field Efield distribution along the cut line of X = 1.9 μm, Z = 6 μm and X = 1.9 μm, Z = 4.9 μm. It is noted that the Efield in the N-drift of the BEA-LDMOS reaches a maximum at the end of the Extended Drain and then drops rapidly, as shown in Figure 9a. This is because the Extended Drain of high doping forms a heterojunction with the drift region of low doping concentration. In heterogeneous junctions, electrons and holes are unevenly distributed on both sides due to different material doping concentrations. Since the concentration of electrons in the highly doped region is higher than that in the low-doped region, electrons will diffuse from the high-doped region to the low-doped region, while holes will diffuse from the low-doped region to the high-doped region. Electrons and holes will meet at the center of the junction, resulting in a large number of recombination. This recombination results in a region of space charges near the junction, and the uneven distribution of charges in this region leads to the formation of a sharp electric field. Meanwhile, the Efield in the buried silicon dioxide layer of the BEA-LDMOS is much higher than those of the other three devices, as shown in Figure 9b. This is because the Extended Drain draws the high electric field in the N-drift region into the buried silicon dioxide layer, which can withstand higher voltages.

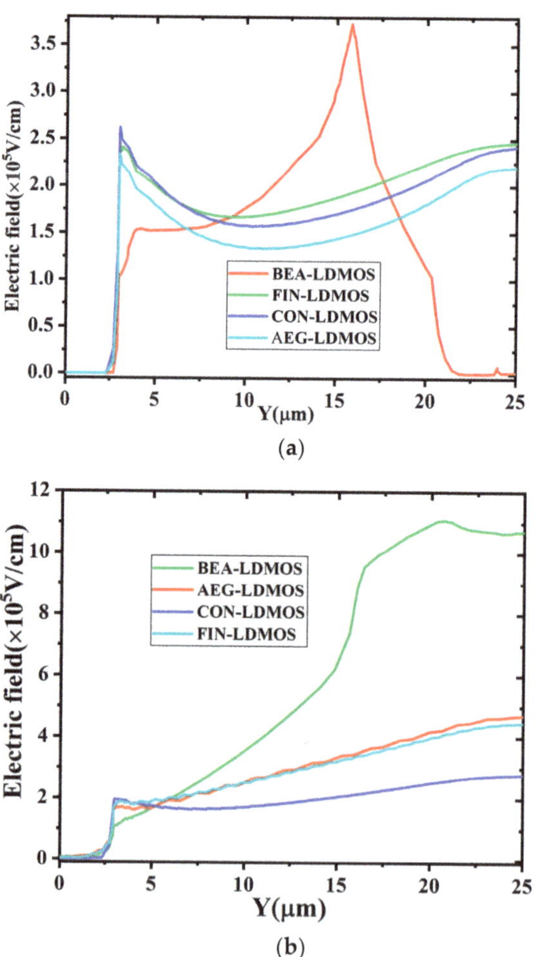

Figure 9. The electric field distribution at the avalanche breakdown at (**a**) X = 1.9 μm, Z = 6 μm; (**b**) X = 1.9 μm, Z = 4.9 μm.

3.3. Influence of Unique Key Parameters on the $R_{on,sp}$, Peak g_m, and BV of the BEA LDMOS

Figure 10 shows the influence of the length (L) and thickness (H) of the Extended Drain on the $R_{on,sp}$ and BV for the BEA-LDMOS. In Figure 10a, the Extended Drain is equivalent to a low-resistance channel, so increasing the thickness of the Extended Drain is equivalent to increasing the volume of the low-resistance channel. Consequently, the specific conduction resistance decreases with increase in thickness of the Extended Drain. In Figure 10b, the specific conduction resistance decreases with increase in the length of the Extended Drain. The BV of the device generally increases first and then decreases with the increase in the thickness and length of the Extended Drain. This is because at the beginning, with the increase in the volume of the Extended Drain, the high electric field can be better introduced into the silicon dioxide layer. However, when the increase exceeds a certain range, it is equivalent to increasing the drift concentration, which will reduce the breakdown voltage. Considering the trade-off property between the $R_{on,sp}$ and BV, the thickness of 0.4 μm and length of 9 μm are selected as the best parameters.

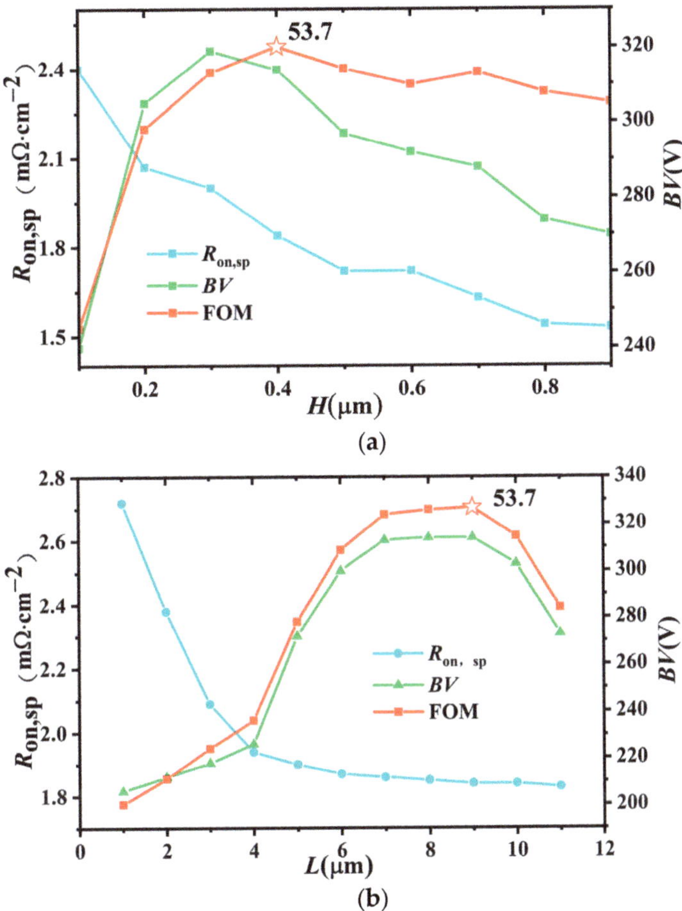

Figure 10. Key parameters: (**a**) thickness (H) of the Extended Drain, (**b**) length (L) influence on the $R_{on,sp}$, peak g_m, and BV for the BEA-LDMOS.

3.4. Dynamic Characteristics

The switching characteristics under inductive load are shown in Figure 11a, and a slower turn-on speed T_{ON} and turn-off speed T_{OFF} of the BEA-LDMOS are observed compared to the CON-LDMOS, AEG-LDMOS, and FIN-LDMOS. Because the gate capacity is proportional to the gate area, the larger the area, the larger the gate capacity. The BEA-LDMOS not only has a FIN structure, but also has an extended superjunction trench gate, so the gate area is much larger than those of the other three devices. Figure 11b shows the effect of different widths of the gate on the switching speed of the device. It can be seen that the wider the gate, the lower the switching speed of the device. The switching performance of the device can be improved by reducing the width of the gate.

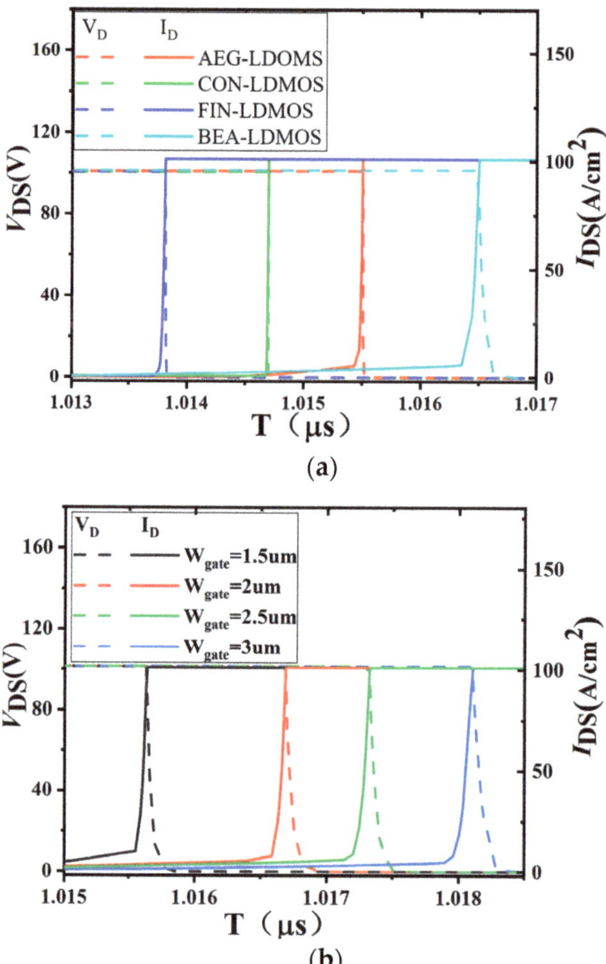

Figure 11. (**a**) Capacitance switching characteristics of the four devices. Turn-on and turn-off curves under inductive load.(**b**) Switching characteristic curves at different gate widths.

3.5. The Trade-Off Property between the $R_{on,sp}$ and BV

Figure 12 demonstrates the trade-off characteristic and FOM for the BEA LDMOS, single RESURF, double RESURF, and triple RESURF in Ref. [9], which are the classic three structures. It can be seen from the figure that the $R_{on,sp}$ of the device is still very low at a larger BV. According to $FOM = BV^2/R_{on,sp}$, it can be concluded that the FOM of BEA is largest and achieves the best trade-off property. The main performance indexes of the four devices compared in this paper are shown in Table 2.

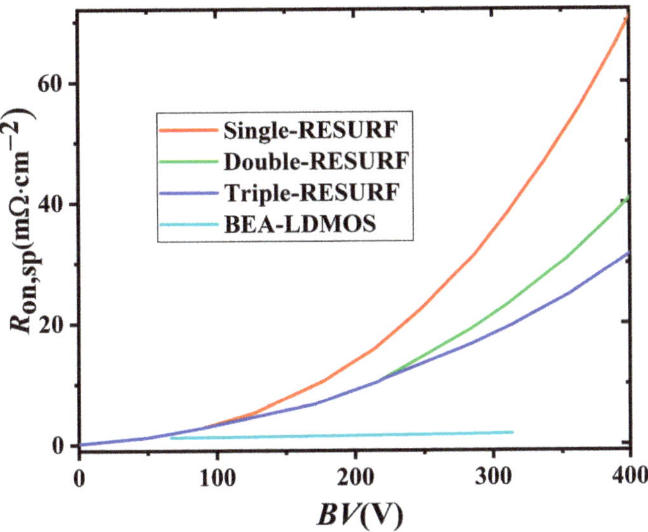

Figure 12. The trade-off relationship between $R_{on,sp}$ and BV for the BEA-LDMOSs and the RESURF. The FOM is calculated by the $FOM = BV^2/R_{on,sp}$.

Table 2. Trade-off Property between the $R_{on,sp}$ and BV.

Symbol	FIN-LDMOS	CON-LDMOS	AEG-LDMOS	BEA-LDMOS
BV (V)	323	329	297	314
$R_{on,sp}$ ($m\Omega \cdot cm^{-2}$)	11.78	14.54	4.41	1.84
FOM	8.86	7.42	20.01	53.43
N_{drift} (cm^{-3})	2.5×10^{15}	2.5×10^{15}	2.5×10^{15}	2.5×10^{15}

4. Conclusions

The mechanism and electric characteristics of the BEA-LDMOS are proposed and researched. The V_{GS} of the BEA is extended through the P-region, and the full buck accumulation effect is formed at the inside of the N-drift, where a 3-D low-resistance channel at the N-drift is achieved. In addition, the Extended Drain is also equivalent to a low-resistance channel. Thus, the $R_{on,sp}$ is significantly decreased. Simultaneously, the superior BV is guaranteed by the charge compensation and assisted depletion effect between the P-type doping and N-drift. Consequently, a FOM of 53.49 MW/cm^2 is achieved, which breaks through the silicon limit of the RESURF.

Author Contributions: Conceptualization, Z.D. and W.C.; methodology, W.C.; software, Z.D.; validation, Z.D., W.C. and H.Z.; formal analysis, Z.D.; investigation, Z.D. and Z.W.; resources, Z.D. and W.C.; data curation, Z.D.; writing—original draft preparation, Z.D.; writing—review and editing, W.C. and Z.W.; visualization, Z.D.; supervision, H.Z.; project administration, Z.H.; funding acquisition, Z.W. All authors have read and agreed to the published version of the manuscript.

Funding: This research received no external funding.

Data Availability Statement: Data is unavailable due to privacy or ethical restrictions.

Conflicts of Interest: The authors declare no conflict of interest.

References

1. Xu, S.; Gan, K.P.; Samudra, G.S.; Liang, Y.C.; Sin, J.K.O. 120 V interdigitated-drain LDMOS (IDLDMOS) on SOI substrate breaking power LDMOS limit. *IEEE Trans. Electron. Devices* **2000**, *47*, 1980–1985. [CrossRef]
2. Park, I.Y.; Choi, Y.I.; Chung, S.K.; Lim, H.J.; Mo, S.I.; Choi, J.S.; Han, M.K. Numerical analysis on the LDMOS with a double epi-layer and trench electrodes. *Microelectron. J.* **2001**, *32*, 497–502. [CrossRef]
3. Erlbacher, T.; Bauer, A.J.; Frey, L. Reduced on Resistance in LDMOS Devices by Integrating Trench Gates into Planar Technology. *IEEE Electron Device Lett.* **2010**, *31*, 464–466. [CrossRef]
4. Zhang, W.; Wang, R.; Cheng, S.; Gu, Y.; Zahng, S.; He, B.; Qiao, M.; Li, Z.; Zhang, B. Optimization and Experiments of Lateral Semi-Superjunction Device Based on Normalized Current-Carrying Capability. *IEEE Electron Device Lett.* **2019**, *40*, 1969–1972. [CrossRef]
5. Hardikar, S.; Tadikonda, R.; Green, D.W.; Vershinin, K.V.; Narayanan, E.M.S. Realizing high-voltage junction isolated LDMOS transistors with variation in lateral doping. *IEEE Trans. Electron. Devices* **2004**, *51*, 2223–2228. [CrossRef]
6. Saxena, R.S.; Kumar, M.J. A new buried-oxide-in-drift-region trench MOSFET with improved breakdown voltage. *IEEE Electron Device Lett.* **2009**, *30*, 990–992. [CrossRef]
7. Qiao, M.; Li, Y.; Zhou, X.; Li, Z.; Zhang, B. A 700- V Junction-Isolated Triple RESURF LDMOS with N-Type Top Layer. *IEEE Electron Device Lett.* **2014**, *35*, 774–776. [CrossRef]
8. Baliga, B.J. Power semiconductor device figure of merit for high-frequency applications. *IEEE Electron Device Lett.* **1989**, *10*, 455–457. [CrossRef]
9. Iqbal, M.M.; Udrea, F.; Napoli, E. On the static performance of the RESURF LDMOSFETS for power ICs. In Proceedings of the International Symposium on Power Semiconductor Devices, Barcelona, Spain, 14–17 June 2009; pp. 247–250. [CrossRef]
10. Ge, W.; Luo, X.; Wu, J.; Lv, M.; Wei, J.; Ma, D.; Deng, G.; Cui, W.; Yang, Y.; Zhu, K. Ultra-Low On-Resistance LDMOS With Multi-Plane Electron Accumulation Layers. *IEEE Electron Device Lett.* **2017**, *38*, 910–913. [CrossRef]
11. Chen, W.; Qin, H.; Huang, Y.; Huang, Y.; Han, Z. A Bulk Full-Gate SOI-LDMOS Device with Bulk Channel and Electron Accumulation Effect. *IEEE Trans. Electron Devices* **2021**, *68*, 6286–6291. [CrossRef]
12. Chen, W.; Qin, H.; Zhang, H.; Han, Z. Bulk Electron Accumulation LDMOS With Extended Superjunction Gate. *IEEE Trans. Electron Devices* **2022**, *69*, 1900–1905. [CrossRef]
13. Guo, Y.; Yao, J.; Zhang, B.; Lin, H.; Zhang, C. Variation of Lateral Width Technique in SoI High-Voltage Lateral Double-Diffused Metal–Oxide–Semiconductor Transistors Using High-k Dielectric. *IEEE Electron Device Lett.* **2015**, *36*, 262–264. [CrossRef]
14. Yao, J.; Zhang, Z.; Guo, Y.; Wu, J.; He, Y.; Li, M.; Lin, H. Novel LDMOS With Integrated Triple Direction High-k Gate and Field Dielectrics. *IEEE Trans. Electron Devices* **2021**, *68*, 3997–4003. [CrossRef]
15. Wang, Z.; Zhang, B.; Fu, Q.; Xie, G.; Li, Z. An L-Shaped Trench SOI-LDMOS With Vertical and Lateral Dielectric Field Enhancement. *IEEE Electron Device Lett.* **2012**, *33*, 703–705. [CrossRef]
16. Zhang, W.; Zhan, Z.; Yu, Y.; Cheng, S.; Gu, Y.; Zhang, S.; Luo, X.; Li, Z.; Qiao, M.; Li, Z.; et al. Novel Superjunction LDMOS (>950 V) With a Thin Layer SOI. *IEEE Electron Device Lett.* **2017**, *38*, 1555–1558. [CrossRef]
17. Zhang, B.; Li, Z.; Hu, S.; Luo, X. Field Enhancement for Dielectric Layer of High-Voltage Devices on Silicon on Insulator. *IEEE Trans. Electron Devices* **2009**, *56*, 2327–2334. [CrossRef]
18. Wei, J.; Zhang, M.; Jiang, H.; Zhou, X.; Li, B.; Chen, K.J. Superjunction MOSFET With Dual Built-In Schottky Diodes for Fast Reverse Recovery: A Numerical Simulation Study. *IEEE Electron Device Lett.* **2019**, *40*, 1155–1158. [CrossRef]
19. Okada, M.; Kyogoku, S.; Kumazawa, T.; Saito, J.; Morimoto, T.; Takei, M.; Harada, S. Superior Short-Circuit Performance of SiC Superjunction MOSFET. In Proceedings of the 2020 32nd International Symposium on Power Semiconductor Devices and ICs (ISPSD), Vienna, Austria, 13–18 September 2020; pp. 70–73. [CrossRef]
20. Udrea, F.; Deboy, G.; Fujihira, T. Superjunction Power Devices, History, Development, and Future Prospects. *IEEE Trans. Electron Devices* **2017**, *64*, 713–727. [CrossRef]
21. Chen, X.B.; Sin, J.K.O. Optimization of the specific on-resistanceof the COOLMOS/sup TM. *IEEE Trans. Electron Devices* **2001**, *48*, 344–348. [CrossRef]
22. Wei, J.; Luo, X.; Zhang, Y.; Li, P.; Zhou, K.; Zhang, B.; Li, Z. High-Voltage Thin-SOI LDMOS With Ultralow ON-Resistance and Even Temperature Characteristic. *IEEE Trans. Electron Devices* **2016**, *63*, 1637–1643. [CrossRef]
23. *Sentaurus Device User Guide*, Version J-2014.09; Synopsys: Mountain View, CA, USA, 2014.
24. Arora, N.D.; Hauser, J.R.; Roulston, D.J. Electron and Hole Mobilities in Silicon as a Function of Concentration and Temperature. *IEEE Trans. Electron Devices* **1982**, *29*, 292–295. [CrossRef]
25. Fossum, J.G.; Mertens, R.P.; Lee, D.S.; Nijs, J.F. Carrier Recombination and Lifetime in Highly Doped Silicon. *Solid-State Electron.* **1983**, *26*, 569–576. [CrossRef]

26. McIntyre, R.J. On the Avalanche Initiation Probability of Avalanche Diodes Above the Breakdown Voltage. *IEEE Trans. Electron Devices* **1973**, *20*, 637–641. [CrossRef]
27. Appels, J.A.; Vaes, H.M.J. High voltage thin layer devices (RESURF devices). In Proceedings of the 1979 International Electron Devices Meeting, Washington, DC, USA, 3–5 December 1979; pp. 238–241. [CrossRef]

 *micromachines*

Article

A False Trigger-Strengthened and Area-Saving Power-Rail Clamp Circuit with High ESD Performance

Boyang Ma [1], Shupeng Chen [1,*], Shulong Wang [1,*], Lingli Qian [2], Zeen Han [1], Wei Huang [1], Xiaojun Fu [3] and Hongxia Liu [1]

1 Key Laboratory of Wide Band-Gap Semiconductor Materials and Devices of Education, The School of Microelectronics, Xidian University, Xi'an 710071, China; 13273711065@163.com (B.M.); hanzeen1212@163.com (Z.H.); huangweifirst@163.com (W.H.); hxliu@mail.xidian.edu.cn (H.L.)
2 Chongqing Acoustic-Optic-Electronic Co., Ltd. of CETC & 24 Institute, Chongqing 401331, China; qlingli@163.com
3 National Key Laboratory of Integrated Circuits and Microsystems, Chongqing 401332, China; xjfu2000@163.com
* Correspondence: spchen@xidian.edu.cn (S.C.); slwang@xidian.edu.cn (S.W.)

Abstract: A power clamp circuit, which has good immunity to false trigger under fast power-on conditions with a 20 ns rising edge, is proposed in this paper. The proposed circuit has a separate detection component and an on-time control component which enable it to distinguish between electrostatic discharge (ESD) events and fast power-on events. As opposed to other on-time control techniques, instead of large resistors or capacitors, which can cause a large occupation of the layout area, we use a capacitive voltage-biased p-channel MOSFET in the on-time control part of the proposed circuit. The capacitive voltage-biased p-channel MOSFET is in the saturation region after the ESD event is detected, which can serve as a large equivalent resistance (~10^6 Ω) in the structure. The proposed power clamp circuit offers several advantages compared to the traditional circuit, such as having at least 70% area savings in the trigger circuit area (30% area savings in the whole circuit area), supporting a power supply ramp time as fast as 20 ns, dissipating the ESD energy more cleanly with little residual charge, and recovering faster from false triggers. The rail clamp circuit also offers robust performance in an industry-standard PVT (process, voltage, and temperature) space and has been verified by the simulation results. Showing good performance of human body model (HBM) endurance and high immunity to false trigger, the proposed power clamp circuit has great potential for application in ESD protection.

Keywords: power clamp circuit; ESD; HBM; false trigger

Citation: Ma, B.; Chen, S.; Wang, S.; Qian, L.; Han, Z.; Huang, W.; Fu, X.; Liu, H. A False Trigger-Strengthened and Area-Saving Power-Rail Clamp Circuit with High ESD Performance. *Micromachines* **2023**, *14*, 1172. https://doi.org/10.3390/mi14061172

Academic Editor: Aiqun Liu

Received: 8 May 2023
Revised: 28 May 2023
Accepted: 30 May 2023
Published: 31 May 2023

1. Introduction

With the development of integrated circuits (ICs), ESD protection has become the major concern regarding the reliability of IC products [1–3]. In order to solve this problem, researchers have proposed the gate-grounded NMOS (GGNMOS), the silicon-controlled rectifier (SCR) structures, and the RC-based power-rail clamp circuit, which can provide a low resistance path to achieving ESD protection without affecting the fragile core circuit [4–9]. Among them, the RC-based power-rail clamp circuit has become the mainstream protection method in full chip ESD protection due to its low trigger voltage and mature manufacturing process [4–8].

However, the RC-based power-rail clamp circuit is often falsely triggered by severe power noise or fast power-up events, which leads to the burning out of the clamp MOSFET, as shown in Figure 1. To improve immunity to false trigger and make sure it has sufficient turn-on time during ESD events, the hybrid triggering method combining static and transient efficiency has been proposed [9,10]. However, the sustained leakage current path in [9] causes massive unnecessary power consumption, while the circuit in [10] shows false trigger in fast power-on pulses with 20 ns rise time. Another improvement method uses the separation technique of a detection component and an on-time control component, which makes the clamping MOSFET turn-on time completely independent of the ESD-transient detection circuit [11–13]. Figure 2a shows the overall circuit structure design method to prevent false trigger, consisting of three parts: the detection component, the on-time control component, and the clamp device. The traditional power clamp circuit separating the detection and on-time control component proposed by Miller et al. is shown in Figure 2b [11], and the modified circuit with a current mirror proposed by Qi Liu et al. is shown in Figure 2c [12]. Both of the two ESD circuits here can achieve voltage clamping and prevent false trigger during fast power-on events. However, the large resistors and capacitors in these two ESD circuits will undoubtedly increase the layout, which can ultimately cause an increase in the manufacturing costs of ICs. Furthermore, there is still the possibility of false trigger in the case of high frequency and large amplitude noise disturbance.

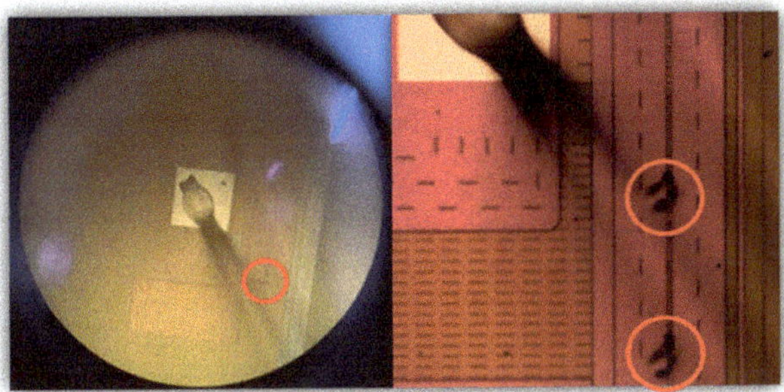

Figure 1. A picture of the burning out of the clamp MOSFET after testing. (The burnt places are circled in red).

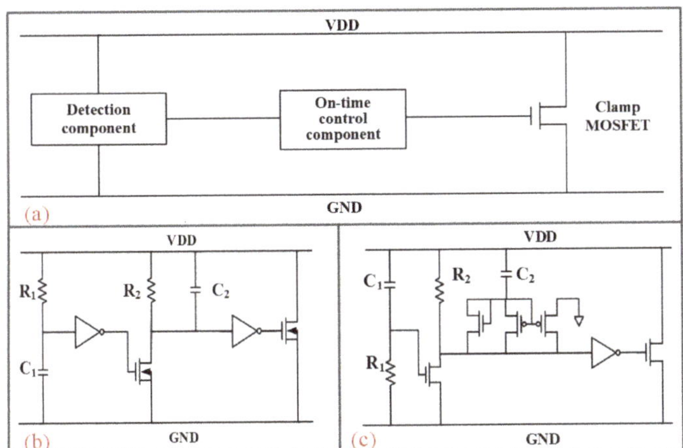

Figure 2. (**a**) Overall structure of a power supply clamp circuit. (**b**) The traditional power clamp circuit with separate detection and on-time control components [11]. (**c**) The modified circuit with current mirror [12].

Therefore, an area-saving power clamp circuit which has good immunity to false trigger is proposed in this paper. As opposed to other on-time control techniques, we use a capacitive voltage-biased p-channel MOSFET in the on-time control part to realize Mega Ohm-level large equivalent impedance. Simulation results verify that the proposed power clamp circuit is area-saving. Furthermore, it can achieve μs level transient turn-on time to fully release ESD stress, while the RC time constant of the detection part is only 10 ns to avoid most of the false trigger events. In addition, the circuit also has a low standby leakage current under normal power-on conditions.

2. The Proposed Power-Rail ESD Clamp Circuit

2.1. Structure of Proposed Circuit

In general, for common HBM ESD signals the rising edge is less than 10 ns [14,15]. Therefore, only a very small time constant (10 ns) is required for the usual ESD signal detection component. However, due to the requirements of the electrostatic discharge time, the RC time constant setting is usually relatively large, close to 1 μs. This causes false triggering during fast power-on events, resulting in an abnormal opening of the discharge circuit, which in turn leads to an increase in chip power consumption or even burnout.

Figure 3 shows the proposed power-rail clamp circuit. Due to the separation of the detection component and the on-time control component, it has a relatively small RC time constant (10 ns) in the ESD detection component, which results in good immunity to false trigger and reduces leakage during power-on, as shown in the purple box in Figure 3a. The on-time control component consists of charging and discharging modules, as shown in the green box in Figure 3a. The charging module is composed of a small capacitor C2 (100 fF) and an nmos2, which is responsible for pulling the node B to a low level by charging the C2. The discharge module is composed of a p-channel MOSFET, C2, and nmos1, which is responsible for pulling the node B to a high level by discharging the C2. Particularly, the capacitance between the gate and drain of the nmos1 and the capacitance between the gate and source of the p-channel MOSFET form a capacitive voltage divider, which can provide gate voltage for the p-channel MOSFET. The equivalent schematic diagram of the capacitive voltage divider circuit is shown in Figure 3b. Combining Equations (1) and (2), the equivalent resistance (r_{ds}) of MOSFET is inversely proportional to the difference between the gate voltage and the source voltage (V_{gs}).

$$I_{ds} = \mu C_{ox}\frac{W}{L}(V_{gs} - V_{th})^2(1 + \lambda V_{ds}) \quad (1)$$

$$r_{ds} = \varphi V_{ds} / \varphi I_{ds} \quad (2)$$

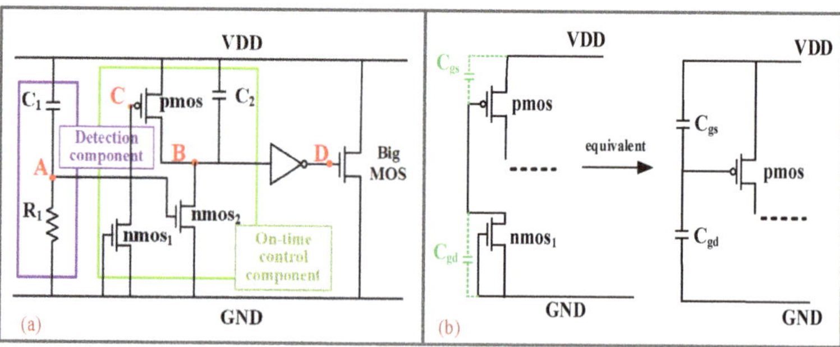

Figure 3. (**a**) The proposed power-rail clamp circuit. (**b**) The equivalent schematic diagram of the capacitive voltage divider circuit.

If the p-channel MOSFET is biased at a relatively high gate voltage, the p-channel MOSFET will be equivalent to a huge resistance (~10^6 Ω) after ESD events are detected in the circuit. In this way, the discharge time of C2 can be prolonged, and the voltage of node B can be raised slowly to ensure that the ESD clamping MOSFET has enough turn-on time to discharge static electricity. Moreover, the leakage current of the on-time control component is negligible (nA level) because the nmos1 and nmos2 are always in the off state after normal power-on.

2.2. Principle of Operation

Figure 4a shows the voltage of the key node under a 1.8 V/100 μs power supply. When the normal power-on happens, the transient detection component will not be triggered due to the fact that there is no rapidly rising ESD signal. So, the voltage of node A is kept at a low level and the nmos2 is kept off all the time. Meanwhile, the p-channel MOSFET is turned on quickly with the gate-biased voltage provided by the capacitive voltage divider. So, the voltage of node B is pulled up to a high level through the p-channel MOSFET and C2. Then the voltage of node D is kept at a level of zero and the clamping MOSFET (the BIGMOS in Figure 3a) is kept off. That is to say, the proposed power clamp circuit can accurately identify the power signal and maintain a standby state while the internal circuit is working normally.

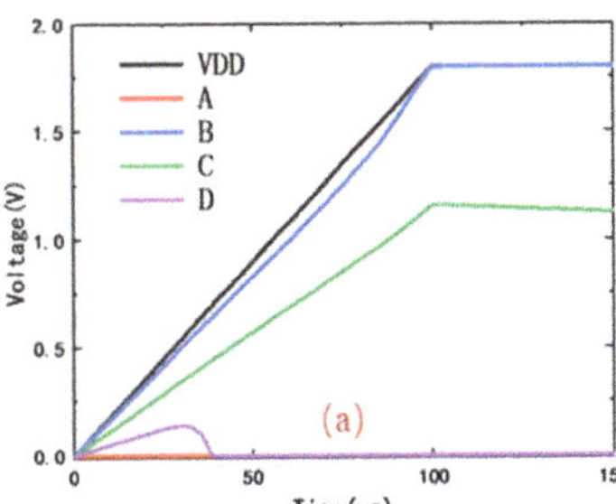

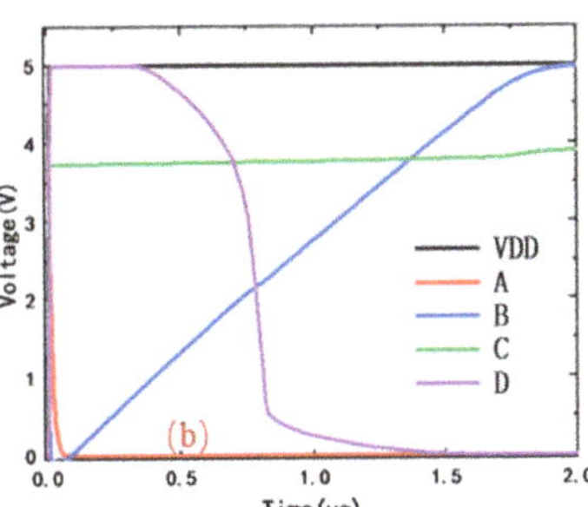

Figure 4. (**a**) The voltage of key node under 1.8 V/100 μs power supply. (**b**) The voltage of key node under ESD events (Simulated by 5 V/10 ns pulse waveform).

Figure 4b shows the voltage of the key node under ESD events. When the ESD events happen, the transient detection component with a Rl· C1 time constant is triggered by an ESD signal, and the voltage of node A is pulled up to a high level to turn on the nmos2. At this moment, the p-channel MOSFET is turned on quickly with the gate-biased voltage provided by the capacitive voltage divider. Then the voltage of node B is pulled down to a low level and the voltage of node D is pulled up to a high level. So, the clamping MOSFET is turned on.

A little time later (proportional to Rl· C1 time constant), the voltage of node A changes to a low level, and the nmos2 is turned off. Then the C2 is discharged through the p-channel MOSFET. While the p-channel MOSFET in the saturation region is equivalent to a large resistance, because its long channel is narrow and pinched-off it takes a relatively long time for node B to be pulled up to a high level. Finally, the voltage of node D is pulled down to a low level, which turns off the clamping MOSFET. Therefore, the proposed clamp circuit can quickly recognize the ESD signal and ensure sufficient time (μs level) to discharge static electricity.

3. Simulation and Results Discussion

Comprehensive simulation tests were conducted to illustrate the advantage of the proposed structure. All the tests were carried out on a Cadence simulation test platform based on the 180 nm process.

3.1. The Circuit-Level TLP Test

The transient transmission line pulsing test (TLP) is specifically designed to be one of the most effective methods used to verify the protection level of ESD circuits. Here, a square wave with a rising time of 10 ns and a voltage amplitude of 0–5 V was used to simulate the TLP stress [16,17].

Table 1 shows the main parameters used in the traditional circuit, the modified circuit, and the circuit proposed above. To ensure that the ESD detection capabilities of the three structures are the same, we kept R1 and C1, which are 10 kΩ and 1 pf, respectively, consistent in the three circuits. R2 and C2 are the key parameters to control the electrostatic discharge time, depending on the circuit structure used. The width of the clamping MOSFET (W_{mos}) was set to 2000 µm, which can provide the same low resistance path to achieve ESD protection. After applying the same TLP stress to the above three circuits, the test results are shown in Figure 5a. We can clearly see that the discharge times of both traditional and modified circuits are 480 ns and 710 ns, respectively. The proposed circuit increases discharge time to 870 ns, which is far longer than the previous structures. This means that the proposed power clamp circuit can discharge static electricity more fully than the other two structures.

Table 1. The main parameters of the three circuits mentioned above.

	The Classic [11]	The Modified [12]	The Proposed
R1	10 K	10 K	10 K
C1	1 p	1 p	1 p
R2	400 K	30 K	Voltage-biased MOSFET
C2	1 p	1 p	100 f
W_{mos}	2000 µ	2000 µ	2000 µ

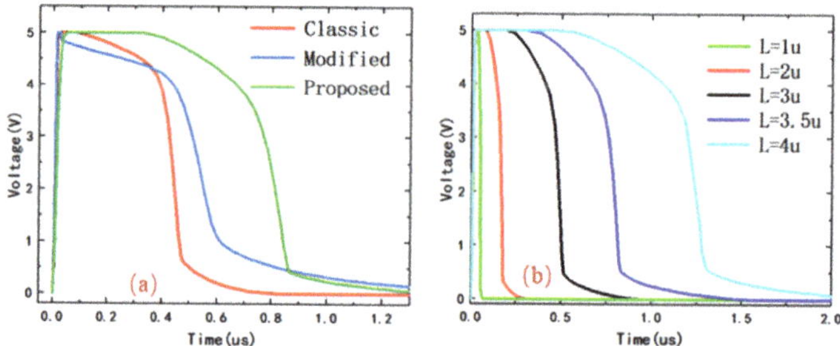

Figure 5. (**a**) The gate voltage of MOSFET of three circuits under circuit-level TLP test. (**b**) The gate voltage of MOSFET varying from the W/L of the p-channel MOSFET during an ESD event.

In general, microsecond electrostatic discharge times are sufficient, but worse conditions, such as surge voltage and current, can occur in different application environments. The proposed voltage clamp circuit can ensure the electrostatic discharge at the microsecond level and realize the adjustable discharge time. Figure 5b shows the gate voltage of the clamping MOSFET varying from the W/L of the p-channel MOSFET during an ESD event. It can be seen that the turn-on time of the clamping MOSFET increases with the increase of channel length. This is because the longer the channel, the larger the equivalent resistance and the longer the discharge time. More importantly, the discharge time can be adjusted by controlling the gate-biased voltage of the p-channel MOSFET, which achieves equivalent resistance adjustability by changing the opening degree of the channel.

3.2. Area-Efficiency Evaluation

Figure 6 depicts layout views of the modified clamp circuits with a current mirror and the proposed circuit mentioned above in which the MOSFET has a default width of 2000 um. The area of the modified clamp circuits with a current mirror is 60 µm × 61 µm in Figure 6a, while the area of the proposed circuit is only 45 µm × 61 µm in Figure 6b. Due to the large equivalent resistance of the voltage-biased p-channel MOSFET in the proposed circuit, it greatly reduces the need for the capacitor C2 while replacing the huge R2 area. Therefore, compared to the modified circuit with a current mirror, at least 30% of the layout area is saved by the proposed circuit.

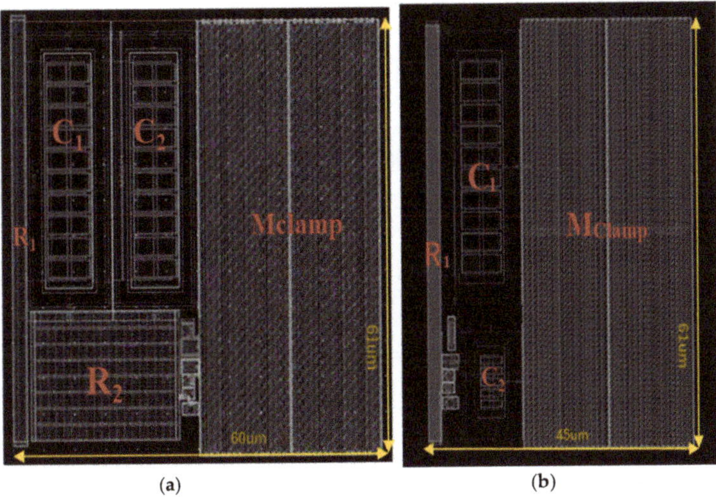

Figure 6. (**a**) Layout of the modified circuit with current mirror. (**b**) Layout of the proposed circuit.

3.3. Circuit-Level ESD Test

The circuit-level ESD test was performed with the HBM to verify the effectiveness of the clamp circuit under ESD conditions. For a 4 KV HBM waveform, the peak current is 2.67 A ± 10% with a rise time of less than 10 ns and a duration of 120–180 ns [18]. The simulation circuit and current waveform of the HBM are shown in Figure 7. In Figure 7a, the C_{esd} and R_{esd} are the equivalent capacitance and equivalent resistance of the human body, respectively. Considering the parasitic effect, we have added capacitance C_p and inductance L_p. The general range of L_p values is 5–12 µH and the range of C_p values is 1–4 pF. The simulated 4 kV HBM waveform is shown in Figure 7b, and it can be seen that the current rise time (t_r) is less than 10 ns, with a peak value (I_p) of approximately 2.67 A, which meets the HBM testing standard.

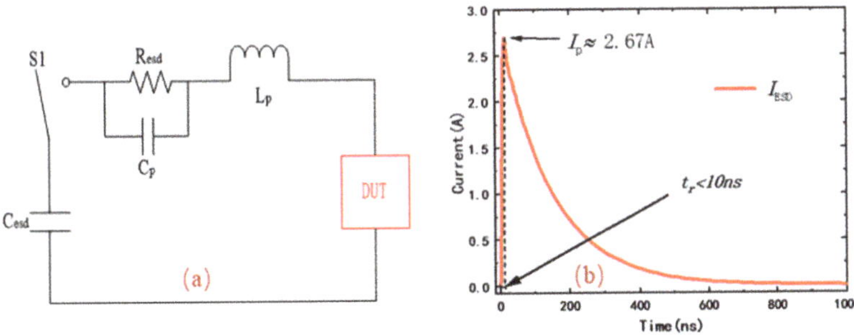

Figure 7. (**a**) The HBM for circuit-level ESD test. (**b**) Simulated HBM 4 kV discharge waveform.

Figure 8 shows the response of VDD under the 4 KV HBM ESD test, and the impact of temperature variations on the proposed power clamp is considered. With the increase of temperature, the maximum clamping voltage of the proposed circuit increases gradually, and the clamping ability of the circuit decreases slightly (the clamping voltage increased from 6.2 V to 6.9 V). This is mainly because the increase in temperature leads to an increase in the equivalent resistance of the clamping MOSFET. On the whole, even under high or low temperature conditions (−40 °C to 125 °C), the proposed circuit can clamp the power supply voltage below 7 V during 4 KV HBM events and can quickly drop to below 4 V within 50 ns. This shows that the proposed circuit has a superior electrostatic clamping ability.

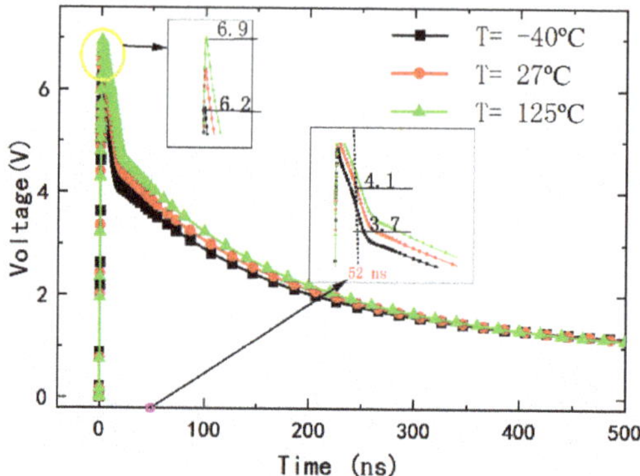

Figure 8. The response of VDD under circuit-level ESD test with 4 KV HBM.

3.4. Immunity to the Fast Power Events

Figure 9 shows the gate voltage of the clamping MOS when fast power-on events happen. The fast power-up events of 1 µs/1.8 V, 200 ns/1.8 V, and 20 ns/1.8 V are simulated, respectively. As we can see in Figure 9, the faster the power supply is powered on, the higher the transient gate voltage of the clamping MOSFET, but the shorter the duration time of the gate voltage. Even for the worst case of a 20 ns/1.8 V fast power-on, the gate voltage of the clamping MOSFET is about 292 mV, far below the threshold voltage (V_{th}). Therefore, the proposed power clamp circuit can effectively avoid false triggering during fast power-on events. This is mainly due to the small time constant (10 ns) of the detection component; the detection circuit can distinguish well between ESD events and fast power-on events.

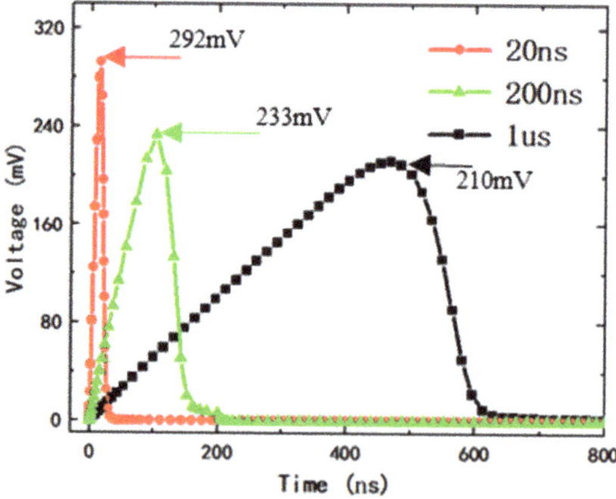

Figure 9. The gate voltage of MOSFET when fast-power-on events happen.

3.5. Immunity to Noise Characteristics

High switching rates usually cause power supply noise, which can cause energy consumption and even falsely trigger the power clamp circuit. So, it is necessary to verify the immunity to power supply noise by simulating a pseudorandom pulse. The added noise has a frequency of 500 MHz and an amplitude of 0.6 V, which is considered the worst case. The noise characteristic of the proposed circuit is shown in Figure 10. It can be seen that the maximum leakage current of the clamping MOSFET is only about 1.2 mA and the duration time is very short. Therefore, the proposed clamp circuit can significantly mitigate high frequency and large amplitude noise disturbance.

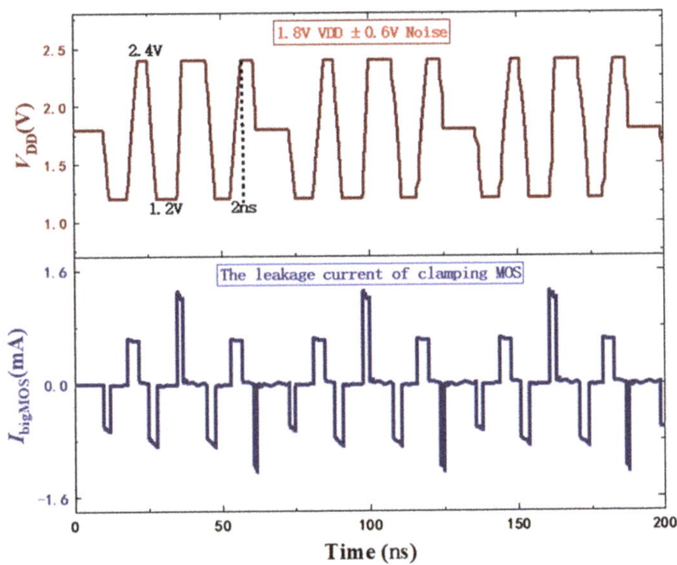

Figure 10. The respond of MOSFET under the disturbance of noise.

3.6. The Low Leakage Characteristic

Figure 11 shows the leakage current of the clamping MOSFET under different temperature conditions. During a fast power-on event when the VDD rises from 0 to 1.8 V (1 µs/1.8 V), the peak leakage current of the clamping MOSFET does not exceed 1.9 µA. After the power-on, the leakage current quickly decreases to the nA level. Although the leakage current slightly increases with the increase of temperature, it is still within an acceptable range (below 60 nA at 125 °C). The results verify that the proposed power clamp circuit is low power, and the energy consumption of the circuit can be ignored after power-on.

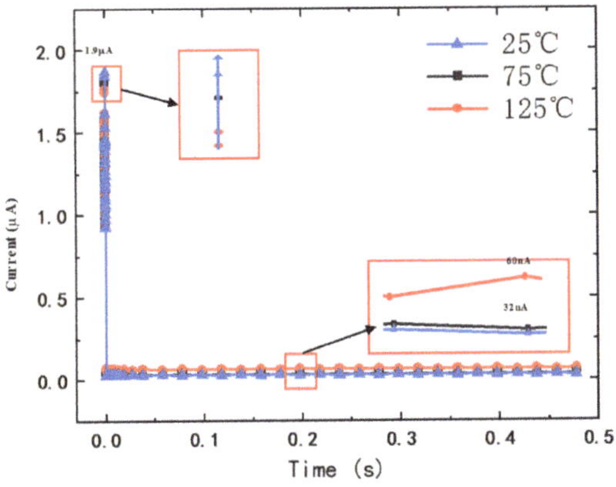

Figure 11. The leakage current under different operating conditions.

4. Performance Comparisons

Table 2 shows the results of comparing the proposed circuit with the other two-stage separated circuits (separate detection component and on-time control component) mentioned above. In the case of ensuring the same ESD detection capability (the same R1 and C1) and electrostatic discharge pathway (the same width of clamping MOSFET (W_{mos})), the proposed circuit occupies the smallest layout area, only 2745 μm^2. At the same time, the proposed circuit has extremely low static leakage and remains at around 31 nA after power-on.

Table 2. The comparison results of the three circuits mentioned above.

	The Classic [11]	The Modified [12]	The Proposed
Process	180 nm	180 nm	180 nm
W_{mos}	2000 μ	2000 μ	2000 μ
Layout area	>5000	3660	2745
I_{leak} at 27 °C	31 nA	μA level	31 nA
False trigger	immune	immune	immune
HBM level	4 KV	4 KV	4 KV

Performance comparisons of the proposed clamp with the representative prior approaches are presented in Table 3. Regarding the trigger circuit (TC) area-reduction ratio in Table 3, the baseline circuit for comparison is the traditional transient circuit with an RC time constant of 100 ns. By using a capacitor-biased p-channel MOSFET to achieve equivalent large resistance, the proposed circuit obtains a higher area efficiency than most prior types. The transient response time of the proposed circuit at the μs level is sufficient for ESD current discharge. Additionally, the circuit demonstrated high false trigger immunity in the worst case of fast power-up pulses. The most important quality of the proposed clamp is the adjustable transient response time which suits various ESD protection scenarios, which is not achieved in existing hybrid triggering clamps. By comparison, the proposed circuit performs better than prior circuits and provides an excellent solution for on-chip ESD protection.

Table 3. Performance comparisons of the proposed clamp with the representative prior clamps.

	TED 2018 [16]	TDMR 2020 [19]	ISCAS 2022 [20]	TED 2022 [21]	The Proposed
Process	180 nm	BCD Process	28 nm	28 nm	180 nm
I_{leak}	N/A	31 nA	7 nA	6.8 nA	31 nA
TC area-reduction ratio	No Reduction	No Reduction	~50% over the baseline circuit	~90% over the baseline circuit	>70% over the baseline circuit
False trigger	immune	immune	immune	immune	immune
Transient Response Time	100 ns	μs—level	μs—level	μs—level	μs—level
Adjustable time	NO	NO	YES	NO	YES

5. Conclusions

A power clamp circuit which has good immunity to false trigger was proposed in this paper. On the one hand, by utilizing the principle of capacitive voltage division, the voltage-biased p-channel MOSFET in the discharge module is equivalent to a huge resistance (~10^6 Ω) after ESD events are detected in the circuit. Thus, microsecond discharge times can be easily achieved while avoiding the use of large resistors and capacitors. Compared with traditional circuits, the proposed circuit area savings is at least 30% (trigger circuit area savings is at least 70%). On the other hand, the proposed circuit has a strong ability to prevent false triggering, supporting a power-on time of as fast as 20 ns and withstanding high-frequency noises of 500 MHZ/0.6 V. In addition, when the internal circuit is working normally, the proposed circuit can maintain a standby state, and the low standby leakage current is only 31 nA, avoiding energy consumption. In ESD events, the clamping MOSFET can be turned on quickly, forming a low resistance path to fully discharge static electricity.

Even under high or low temperature conditions (−40 °C to 125 °C), the proposed circuit can clamp the power supply voltage below 7 V during 4 KV HBM events and can quickly drop to below 4 V within 50 ns. Therefore, the proposed circuit exhibits good HBM endurance and high immunity to false trigger, which have great application potential in ESD protection.

Author Contributions: Conceptualization and writing—original draft preparation, B.M.; methodology and funding acquisition, S.C. and S.W.; validation, L.Q. and Z.H.; writing—review and editing, W.H. and X.F.; funding acquisition, H.L. All authors have read and agreed to the published version of the manuscript.

Funding: This research is financially supported by the Open Foundation of National Key Laboratory of Integrated Circuits and Microsystems (Grant No. JCKY2022210C006), Chongqing Acoustic-optic-electronic Co.Ltd. of CETC & 24 Institute Project (Grant No. WJS2402202208002J), National Natural Science Foundation of China (Grant No. U2241221).

Data Availability Statement: Data are contained within the article.

Conflicts of Interest: We declare that we do not have any commercial or associative interest that represents a conflict of interest in connection with the submitted work.

References

1. Mergens, M.; Wybo, G.; Van Camp, B.; Keppens, B.; De Ranter, F.; Verhaege, K.; Jozwiak, P.; Armer, J.; Russ, C. ESD protection circuit design for ultra-sensitive IO applications in advanced sub-90 nm CMOS technolo-gies. In Proceedings of the 2005 IEEE International Symposium on Circuits and Systems, Kobe, Japan, 23–26 May 2005; pp. 1194–1197.
2. Duvvury, C. ESD qualification changes for 45nm and beyond. In Proceedings of the 2008 IEEE International Electron Devices Meeting, San Francisco, CA, USA, 15–17 December 2008; pp. 1–4. [CrossRef]
3. Ker, M.-D. Whole-chip ESD protection design with efficient VDD-to-VSS ESD clamp circuits for submicron CMOS VLSI. *IEEE Trans. Electron Devices* **1999**, *46*, 173–183. [CrossRef]
4. Karalkar, S.P.; Ganesan, V.; Paul, M.; Hwang, K.; Gauthier, R. Design Optimization of MV-NMOS to Improve Holding Voltage of a 28 nm CMOS Technology ESD Power Clamp. In Proceedings of the 2021 IEEE International Reliability Physics Symposium (IRPS), Monterey, CA, USA, 21–24 May 2021; pp. 1–5. [CrossRef]
5. Li, C.; Zhang, F.; Wang, C.; Qi, C.; Lu, F.; Wang, H.; Di, M.; Cheng, Y.; Zhao, H.; Wang, A. Temperature Dependence of Diode and ggNMOS ESD Protection Structures in 28 nm CMOS. In Proceedings of the 2018 14th IEEE International Conference on Solid-State and Integrated Circuit Technology (ICSICT), Qingdao, China, 31 October–3 November 2018; pp. 1–3. [CrossRef]
6. Song, W.; Du, F.; Hou, F.; Liu, J.; Huang, X.; Liu, Z.; Liou, J.J. Design of A Novel Low Voltage Triggered Silicon Controlled Rectifier (SCR) for ESD Applications. In Proceedings of the 2020 International EOS/ESD Symposium on Design and System (IEDS), Chengdu, China, 23–25 June 2021. [CrossRef]
7. Qian, L.; Li, M.; Wang, Y.; Wu, H.; Liu, T.; Guo, J.; Zhu, W.; Hu, S. A Novel Segmented LDMOS-SCR Structure With 8-kV HBM ESD Robustness in CMOS Analog Multiplexer. *IEEE Trans. Electron Devices* **2022**, *69*, 6904–6909. [CrossRef]
8. Di, M.; Pan, Z.; Li, C.; Wang, A. ESD Design Verification Aided by Mixed-Mode Multiple-Stimuli ESD Simulation. *IEEE J. Electron Devices Soc.* **2021**, *9*, 1194–1201. [CrossRef]
9. Lu, G.; Wang, Y.; Wang, Y.; Zhang, X. Low-Leakage ESD Power Clamp Design with Adjustable Triggering Voltage for Nanoscale Applications. *IEEE Trans. Electron Devices* **2017**, *64*, 3569–3575. [CrossRef]
10. Lu, G.; Wang, Y.; Zhang, X. Transient and Static Hybrid-Triggered Active Clamp Design for Power-Rail ESD Protection. *IEEE Trans. Electron Devices* **2016**, *63*, 4654–4660. [CrossRef]
11. Miller, J.W.; Torres, C.A.; Cooper, T.L. Circuit for Electrostatic Discharge Protection. U.S. Patent No. 5946177, 31 August 1999.
12. Liu, Q.; Wang, Y.; Cao, J.; Lu, G.; Zhang, X. Design of power-rail ESD clamp circuit with separate detection component and delay component against mis-triggering. In Proceedings of the 2013 IEEE International Conference of Electron Devices and Solid-State Circuits, Singapore, 1–4 June 2015; pp. 1–3. [CrossRef]
13. Cao, J.; Xue, X.; Wang, Y.; Lu, G.; Zhang, X. A prolonged discharge time ESD power-rail clamp circuit structure with strong ability to prevent false triggering. In Proceedings of the 2016 13th IEEE International Conference on Solid-State and Integrated Circuit Technology (ICSICT), Hangzhou, China, 25–28 October 2016; pp. 1291–1293. [CrossRef]
14. Wang, A. ESDTest Models and Standards. In *Practical ESD Protection Design*; Wiley-IEEE Press: Hoboken, NJ, USA, 2022; pp. 51–76. [CrossRef]
15. *EOS/ESD Standard for ESD Sensitivity Testing*; EOS/ESD Association: New York, NY, USA, 1993.
16. Chen, J.-T.; Ker, M.-D. Design of Power-Rail ESD Clamp With Dynamic Timing-Voltage Detection Against False Trigger During Fast Power-ON Events. *IEEE Trans. Electron Devices* **2018**, *65*, 838–846. [CrossRef]

17. Cao, J.; Ye, Z.; Wang, Y.; Lu, G.; Zhang, X. A Low Leakage Power Clamp ESD Protection Circuit with Prolonged ESD Dis-charge Time and Compact Detection Network. In Proceedings of the 2015 IEEE 11th International Conference on ASIC (ASICON), Chengdu, China, 3–6 November 2015.
18. Liu, H.; Yang, Z.; Li, L.; Zhuo, Q. A novel ESD power supply clamp circuit with double pull-down paths. *Sci. China Inf. Sci.* **2012**, *56*, 1–8. [CrossRef]
19. Narita, K.; Okushima, M. Low-Leakage and Variable VHOLD Power Clamp for Wide Stress Time Range From ESD to Surge Test. *IEEE Trans. Device Mater. Reliab.* **2020**, *20*, 641–649. [CrossRef]
20. Shen, Z.; Wang, Y.; Zhang, X.; Wang, Y. A Novel Low-Leakage ESD Power Clamp Circuit with Adjustable Transient Response Time. In Proceedings of the 2022 IEEE International Symposium on Circuits and Systems (ISCAS), Austin, TX, USA, 27 May–1 June 2022; pp. 3488–3492. [CrossRef]
21. Shen, Z.; Wang, Y.; Zhang, X.; Wang, Y. Area-Efficient Power-Rail ESD Clamp Circuit With False-Trigger Immunity in 28nm CMOS Process. *IEEE J. Electron Devices Soc.* **2022**, *10*, 876–884. [CrossRef]

micromachines

Article

Enhanced Operational Characteristics Attained by Applying HfO_2 as Passivation in AlGaN/GaN High-Electron-Mobility Transistors: A Simulation Study

Jun-Hyeok Choi [1], Woo-Seok Kang [1], Dohyung Kim [1], Ji-Hun Kim [1], Jun-Ho Lee [1], Kyeong-Yong Kim [1], Byoung-Gue Min [2], Dong Min Kang [2] and Hyun-Seok Kim [1,*]

[1] Division of Electronics and Electrical Engineering, Dongguk University-Seoul, Seoul 04620, Republic of Korea; junhyeok6293@dgu.ac.kr (J.-H.C.); kws1117@dongguk.edu (W.-S.K.); ehgudakr@dongguk.edu (D.K.); kjsuk0105@dongguk.edu (J.-H.K.); steve1211@dgu.ac.kr (J.-H.L.); kky0213@dongguk.edu (K.-Y.K.)

[2] Electronics and Telecommunications Research Institute, Daejeon 34129, Republic of Korea; minbg@etri.re.kr (B.-G.M.); kdm1597@etri.re.kr (D.M.K.)

* Correspondence: hyunseokk@dongguk.edu; Tel.: +82-2-2260-3996

Abstract: This study investigates the operating characteristics of AlGaN/GaN high-electron-mobility transistors (HEMTs) by applying HfO_2 as the passivation layer. Before analyzing HEMTs with various passivation structures, modeling parameters were derived from the measured data of fabricated HEMT with Si_3N_4 passivation to ensure the reliability of the simulation. Subsequently, we proposed new structures by dividing the single Si_3N_4 passivation into a bilayer (first and second) and applying HfO_2 to the bilayer and first passivation layer only. Ultimately, we analyzed and compared the operational characteristics of the HEMTs considering the basic Si_3N_4, only HfO_2, and HfO_2/Si_3N_4 (hybrid) as passivation layers. The breakdown voltage of the AlGaN/GaN HEMT having only HfO_2 passivation was improved by up to 19%, compared to the basic Si_3N_4 passivation structure, but the frequency characteristics deteriorated. In order to compensate for the degraded RF characteristics, we modified the second Si_3N_4 passivation thickness of the hybrid passivation structure from 150 nm to 450 nm. We confirmed that the hybrid passivation structure with 350-nm-thick second Si_3N_4 passivation not only improves the breakdown voltage by 15% but also secures RF performance. Consequently, Johnson's figure-of-merit, which is commonly used to judge RF performance, was improved by up to 5% compared to the basic Si_3N_4 passivation structure.

Keywords: AlGaN/GaN; high-electron-mobility transistor; passivation; HfO_2

Citation: Choi, J.-H.; Kang, W.-S.; Kim, D.; Kim, J.-H.; Lee, J.-H.; Kim, K.-Y.; Min, B.-G.; Kang, D.M.; Kim, H.-S. Enhanced Operational Characteristics Attained by Applying HfO_2 as Passivation in AlGaN/GaN High-Electron-Mobility Transistors: A Simulation Study. *Micromachines* **2023**, *14*, 1101. https://doi.org/10.3390/mi14061101

Academic Editors: Zeheng Wang and Jingkai Huang

Received: 27 April 2023
Revised: 22 May 2023
Accepted: 22 May 2023
Published: 23 May 2023

1. Introduction

Generally, AlGaN/GaN high-electron-mobility transistors (HEMTs) are widely adopted in power electronics because of their outstanding electronic and material properties, such as high-critical electric field (~3.3 MV/cm) and wide energy bandgap (3.4 eV). Interestingly, these remarkable characteristics make GaN more practicable for high-power and high-frequency applications compared to other materials [1]. Hence, due to these material characteristics, AlGaN/GaN HEMTs exhibit high electron saturation velocity as well as high current density, high thermal reliability, and high breakdown voltage (V_{BD}) [2–4]. In addition, HEMTs based on the AlGaN/GaN heterostructure show admirable performances via a two-dimensional electron gas (2-DEG) in the channel generated by the spontaneous and piezoelectric polarization effects [5,6]. Nevertheless, to sufficiently satisfy the market requirements, GaN-based HEMTs require further research for high-voltage and high-frequency applications [7–9]. It has been demonstrated that the field-plate structures in GaN-based HEMTs are commonly used to increase the V_{BD}, resulting in operational stability and reliability. However, the frequency characteristics are degraded due to the increase in parasitic capacitances, such as the gate-to-source capacitance (C_{gs}) and gate-to-drain capacitance (C_{gd}) [10,11]. This clearly shows the advantages and disadvantages of applying field-plates in GaN-based HEMTs due to trade-off between DC and RF characteristics. Additionally, many studies are being conducted to improve the devices' performance [12,13]. As an alternative to HEMTs with field-plate structure, we employed HfO_2 as the passivation to enhance V_{BD}. Interestingly, HfO_2 has a high dielectric constant (~25) and large bandgap energy (5.7 eV), both of which may be exploited to improve the devices' performance in comparison with the basic GaN-based HEMTs with Si_3N_4 passivation [14]. Based on these material properties, it is anticipated that the leakage current and V_{BD} characteristics can be improved. However, HfO_2 passivation in HEMTs also produces additional parasitic capacitances, which may degrade their frequency characteristics. Thus, we suggested the additional structures to secure RF performance while applying HfO_2 as a passivation layer.

In this article, we compare and analyze three different passivation structures which use the basic Si_3N_4, only HfO_2, and HfO_2/Si_3N_4 (hybrid), respectively, as passivation materials. Compared to the basic Si_3N_4 passivation structure, we confirmed that the V_{BD} of the HfO_2 passivation structure improved by approximately 18.8%, but its frequency characteristics were significantly degraded. Meanwhile, the hybrid passivation structure exhibited a slightly reduced V_{BD}, but its frequency characteristics were improved to approximately twice that of the HfO_2 passivation structure. Thus, we optimized the second Si_3N_4 passivation thickness in the hybrid passivation structure to further increase its RF performance. Consequently, the various passivation structures in terms of V_{BD}, on-resistance (R_{on}), and cut-off frequency (f_T) were evaluated using the standard lateral figure-of-merit (LFOM) (=V_{BD}^2/R_{on}) and Johnson's figure-of-merit (JFOM) (=$f_T \times V_{BD}$) [15–17].

2. Materials and Methods

To obtain a reasonable simulation criterion, we first analyzed the fabricated HEMT with a 0.15-μm planar-gate structure [18]. The AlGaN/GaN HEMTs were grown on a 4-inch SiC substrate by using metal–organic chemical vapor deposition. More precisely, the epitaxial layers were composed of a 0.2-μm-thick nucleation layer, a Fe-doped 2-μm-thick GaN buffer layer, and a 25-nm-thick $Al_{0.25}Ga_{0.75}N$ barrier layer. Additionally, the Ohmic metallization of the device was formed by Ti/Al/Ni/Au evaporation followed by rapid thermal annealing at 775 °C for 30 s, and device isolation was achieved by P^+ ion implantation. Next, a 50-nm-thick 1st Si_3N_4 passivation layer was deposited by using plasma-enhanced chemical vapor deposition (PECVD). The first metal interconnections with the Ohmic contacts were formed by the Ti/Au evaporation after etching the 1st Si_3N_4 passivation layer. Further, a planar gate was formed by using single-layer electron beam lithography. More precisely, a gate foot length of 0.15 μm was obtained by electron-beam exposure using poly methyl methacrylate resist, and the 1st Si_3N_4 passivation layer underneath the gate pattern was removed by inductively coupled plasma dry etching. Ni/Au planar-gate metal stack was deposited by electron-beam evaporation and lift-off processes. After this, a 250-nm-thick 2nd Si_3N_4 PECVD film was deposited for device passivation. A source-connected field-plate was formed by using the Ti/Au metal and lift-off process. Finally, the wafer-thinning and backside via-hole process was performed. The scanning electron microscope (SEM) image of the fabricated planar gate AlGaN/GaN HEMT is shown in Figure 1a.

Figure 1b shows the schematic diagram of the basic Si_3N_4 passivation structure of the HEMT. Based on the fabricated device, we determined the structural and material parameters to be utilized for modeling without any other structural changes, such as changes to the planar-gate electrode structure, and while retaining the same gate foot-length of 0.15 μm, including the epitaxial layer. Table 1 provides the specific geometrical parameter information of the basic Si_3N_4 passivation structure used in the simulation.

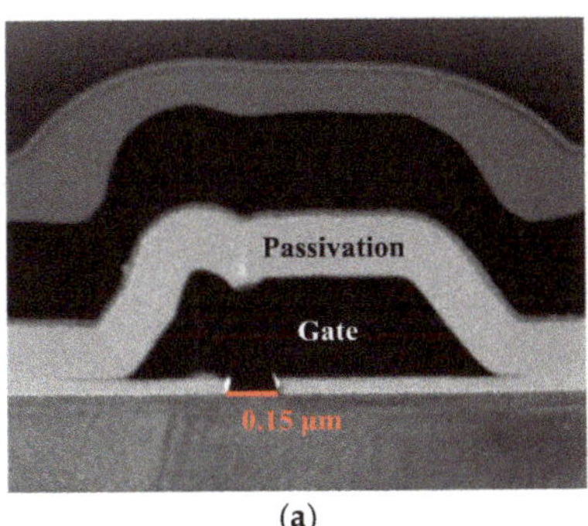

(a)

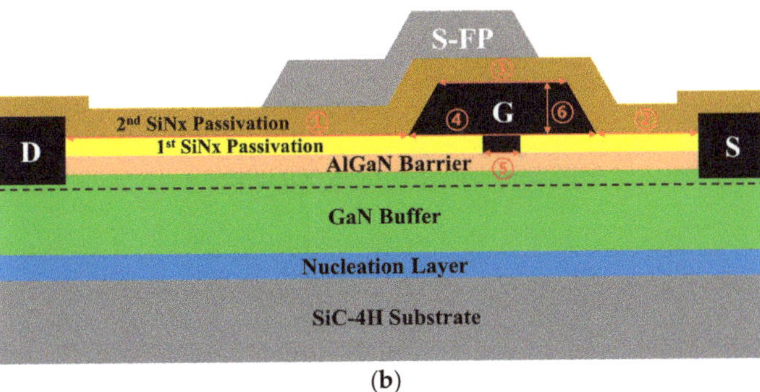

(b)

Figure 1. A cross-sectional schematic of the fabricated planar gate AlGaN/GaN high-electron-mobility transistor (HEMT) structure: (**a**) scanning electron microscope (SEM) image; and (**b**) an illustration used in modeling. The S, D, G, and S-FP represent the source, drain, gate, and source-connected field-plate, respectively; each number (1–6) is explained in Table 1.

Table 1. The geometric parameters of the basic Si_3N_4 passivation structure of HEMT.

Parameters	Value (μm)
① $L_{Gate-Drain}$	1.5
② $L_{Gate-Source}$	0.5
③ $L_{Gate-Head\ top}$	0.8
④ $L_{Gate-Head\ bottom}$	1.0
⑤ $L_{Gate-Foot}$	0.15
⑥ $L_{Gate-Height}$	0.44
Field-plate thickness	0.44
1st passivation	0.05
2nd passivation	0.25
AlGaN barrier	0.025
GaN buffer	2
Nucleation layer	0.2

In this simulation study, it is essential to initialize the material and simulation parameters in order to accurately confirm the operating characteristics of the device. The specific material parameters of GaN and AlGaN used for simulation are summarized in Table 2. As shown in Table 2, we subdivided the FMCT (Farahmand-modified Caughey–Thomas) and GANSAT electron mobility models based on the electric field within the device [19]. Additionally, heat models were applied in the simulation to implement the actual device performance for accurate simulation results. Additionally, acceptor-trap doping in AlGaN/GaN HEMTs is generally used to improve the V_{BD} by reducing the substrate leakage current [20]. However, current-collapse phenomena such as drain-lag and gate-lag inevitably occur [21]. Therefore, a properly controlled acceptor-trap doping is essential to achieve high-performance HEMTs. The Gaussian acceptor doping profile is applied in the simulation by using Fe (iron). More precisely, the peak acceptor-trap doping concentration is set to $10^{18}/cm^2$ in the GaN buffer layer and gradually decreases according to the Gaussian distribution, resulting in an acceptor-trap doping concentration below $10^{15}/cm^2$ at the interface between AlGaN and GaN.

Table 2. Material parameters used in the simulation at a temperature of 300 K (SRH: Shockley–Read–Hall).

Parameters	Units	GaN	AlGaN
Bandgap energy	eV	3.39	3.88
Electron affinity	eV	4.2	2.3
Relative permittivity	-	9.5	9.38
Low field mobility model	-	FMCT Mobility Model	
High field mobility model	-	GANSAT Mobility Model	
Electron saturation velocity	cm/s	1.9×10^7	1.12×10^7
Hole saturation velocity	cm/s	1.9×10^7	1.00×10^6
Electron SRH lifetime	s	1.0×10^{-8}	1.0×10^{-8}
Hole SRH lifetime	s	1.0×10^{-8}	1.0×10^{-8}

In order to conduct an accurate device simulation by considering the self-heating effect (SHE), we applied physical models to calculate the heat generation within the device [22,23]. First, we used the lattice heat flow equation,

$$C\frac{\partial T_L}{\partial t} = \nabla(\kappa \nabla T_L) + H \quad (1)$$

where C is the heat capacitance per unit volume, κ is the thermal conductivity coefficient, H is the heat generation, and T_L is the local lattice temperature. More precisely, the thermal conductivity, which is important to calculate the SHE in a device simulation, is commonly temperature-dependent. Therefore, we applied the thermal conductivity model,

$$\kappa(T) = (TC.CONST)/(T_L/300)^{TC.NPOW} \quad (2)$$

where TC.CONST is the thermal conductivity constant at 300 K and TC.NPOW is the calibration factor which is an experimental value. The applied TC.CONST and TC.NPOW parameters of GaN, AlGaN, and SiC-4H are summarized in the Table 3 [24].

Table 3. Thermal parameters used for the thermal conductivity model.

Parameters	Units	GaN	AlGaN	SiC-4H
TC.CONST	-	1.3	0.4	3.3
TC.NPOW	-	0.43	0	1.61

We investigated the relationship between parasitic capacitances and frequency characteristics. The capacitance equation can be expressed by:

$$C = \frac{\varepsilon_0 \varepsilon_r}{d} A \quad (3)$$

where C is the capacitance, ε_0 is the permittivity of free space (constant value), ε_r is the dielectric constant of the material, A is the area of overlap of the two electrodes, and d is the electrode separation distance. As expressed in Equation (3), ε_r and d have a significant influence on the change in capacitance.

Next, f_T and maximum oscillation frequency (f_{max}) were explained by Equations (4) and (5):

$$f_T = \frac{g_m}{2\pi\left(C_{gs} + C_{gd}\right)} \approx \frac{g_m}{2\pi C_{gs}} \quad (4)$$

$$f_{max} = \frac{f_T}{2\sqrt{\pi f_T C_{gd}\left(R_s + R_g + R_{gs} + 2\pi L_s\right) + G_{ds}\left(R_s + R_g + R_{gs} + \pi f_T L_s\right)}} \approx \sqrt{\frac{f_T}{8\pi R_g C_{gd}}} \quad (5)$$

where g_m, C_{gs}, and C_{gd} represent the transconductance, gate-to-source capacitance, and gate-to-drain capacitance, respectively. As described in Equation (4), decreasing the parasitic capacitances, such as C_{gs} and C_{gd}, increases the f_T. The R_s, R_g, R_{gs}, L_s, and G_{ds} are the source resistance, gate resistance, gate-to-source resistance, parasitic source inductance, and output conductance, respectively [25]. Equation (5) shows that R_g and C_{gd} must be reduced to achieve a higher f_{max}. Additionally, as f_T increases, f_{max} also increases, as shown in Equation (5).

3. Results

3.1. Basic Si_3N_4 Passivation Structure of HEMT Modeling Verified by Matching the Simulation's Results with the Measured Data

In this work, we matched the simulated drain current-gate voltage (I_{DS}-V_{GS}) transfer and f_T with the measured data of the fabricated basic Si_3N_4 passivation structure of the HEMT to ensure the simulation's reliability. The measured datum of the drain current at a gate voltage of 0 V (I_{dss}) was 898.71 mA/mm, which was similar to the simulated datum of 914.90 mA/mm. Furthermore, the measured maximum transconductance (G_m) was 344.17 mS/mm, which corresponds to the simulated value of 349.60 mS/mm. The above results for maintaining the threshold voltage (V_{th}) at −3.8 V were confirmed. Therefore, by adjusting the simulation's parameters, I_{dss}, G_m, and V_{th} values of the simulation and the corresponding measured results were matched within 1.8% of the maximum error rate, as shown in Figure 2a. A dip of the simulated transconductance around-gate voltage of −2.4 V was found, since two different field-dependent electron mobility models were used, as represented in Table 2. The exact criterion for determining the field within the device as low or high remains unknown, but a slight dip in simulated transconductance can occur at an obscure boundary of these models. The I_{DS}–V_{GS} transfer of the fabricated device was measured by using a Cascade Microtech Summit 12,000 probe station and a HP4142B Modular DC Source/Monitor probe station.

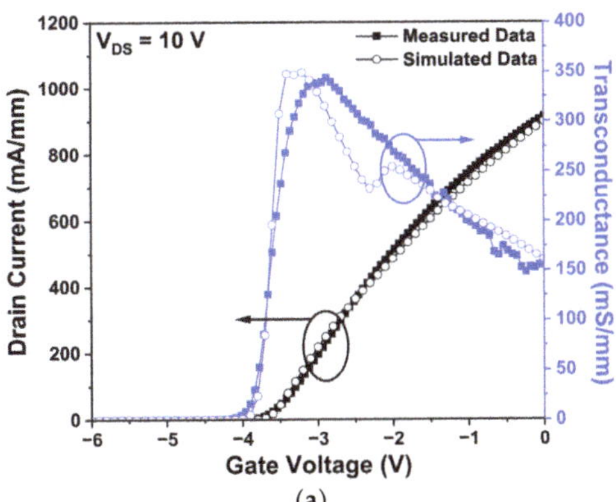

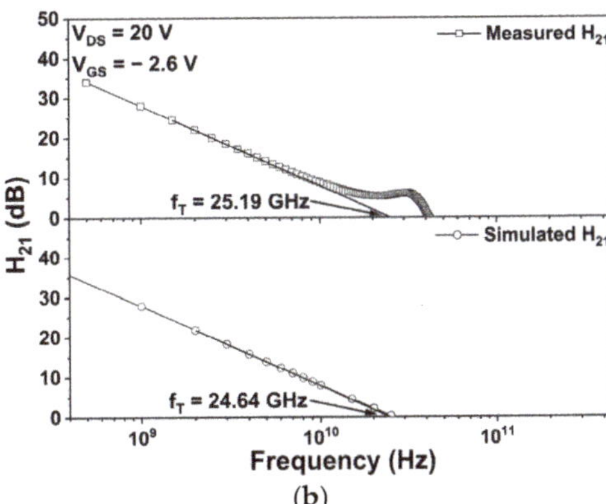

Figure 2. (**a**) The measured and simulated drain current-gate voltage (I_{DS}–V_{GS}) transfer characteristics of a basic Si_3N_4 passivation structure of the HEMT at a drain voltage (V_{DS}) of 10 V. The black and blue arrows represent drain current and transconductance, respectively; and (**b**) the measured and simulated current gain of a basic Si_3N_4 passivation structure of the HEMT as a function of frequency at V_{DS} = 20 V and gate voltage (V_{GS}) = −2.6 V.

The simulated and measured f_T of the basic Si_3N_4 passivation structure of HEMT are shown in Figure 2b. As regards the RF characteristics, the bias points of the simulated results and the measured data were a drain voltage of 20 V and gate voltage of −2.6 V, which were selected since the frequency characteristics were outstanding in comparison to other bias points. More specifically, the f_T was defined as the intersection of the x-axis and the extension line at the point of current gain (H_{21}), with a slope of −20 dB/decade [26]. The measured and simulated values of the f_T were 25.19 GHz and 24.64 GHz, respectively. This clearly shows that the simulated f_T was accurate enough when compared to the measured values, as the error rate was only 2.2%. PNA-X N5245A network analyzer was used to analyze the f_T of the device within the frequency range from 0.5 to 50 GHz.

3.2. Comparative Analysis between HEMTs with Si_3N_4, HfO_2, and Hybrid Passivation Structures

To enhance the operational characteristics, we suggested two structures, as shown in Figure 3. Figure 3a shows the HfO_2 passivation structure of the HEMT. As shown in Figure 3b, the hybrid passivation structure consists of first and second passivation layers, which are composed of HfO_2 and Si_3N_4, respectively. Specifically, these passivation structures will exhibit enhanced DC characteristics, including the V_{BD}, as compared to the basic Si_3N_4 passivation structure, because of the material properties of HfO_2. The other structural parameters excluding the passivation material were not changed in the simulation.

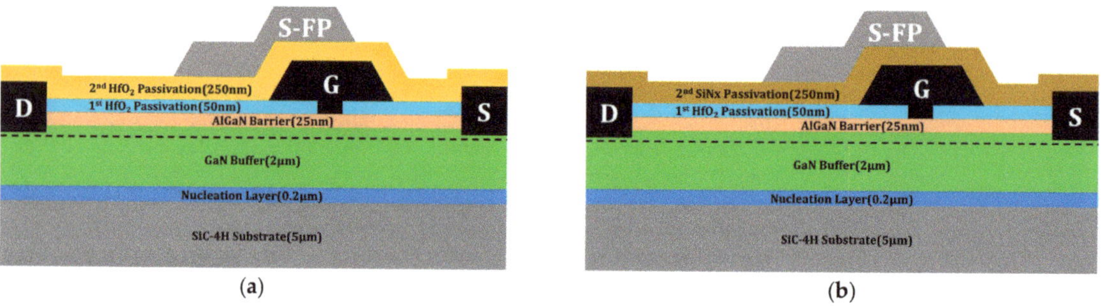

Figure 3. The schematics of various passivation structures for the AlGaN/GaN HEMT: (**a**) HfO_2 passivation structure; and (**b**) hybrid passivation structure.

3.2.1. Analysis of DC Characteristics

First, we analyzed the DC characteristics of the HfO_2 and hybrid passivation-structures, and then compared them to the basic Si_3N_4 passivation-structure. Figure 4a shows the I_{DS}–V_{GS} transfer characteristics of all three structures at a drain voltage of 10 V. Among them, the HfO_2 passivation structure slightly improved not only the drain current, but also the transconductance, in comparison with the basic Si_3N_4 passivation structure. Interestingly, these results show that R_{on} decreases as HfO_2 is employed in passivation [27]. The drain current-drain voltage (I_{DS}–V_{DS}) characteristics were simulated at the gate voltages of −5, −4, −3, −2, −1, and 0 V, respectively, as shown in Figure 4b. As the higher gate voltage was applied, the electron concentration in the channel region increased, resulting in a large drain current. However, a decrease in drain current was observed as the drain voltage increased. These results may be explained by SHE, since applying a higher voltage leads to a higher heat generation, resulting in the degradation of the DC characteristics [28–30]. When the applied drain voltage increased, a strong electric field was generated within the device. Due to the large electric field, phonon scattering was observed to reduce the electron mobility and current density. Although the SHE occurred in all three structures, the HfO_2 passivation and hybrid passivation structures exhibited a higher drain current than did the basic Si_3N_4 passivation structure. In addition, R_{on} was calculated to be 4.02, 3.84, and 3.97 Ω-mm for the basic Si_3N_4, HfO_2, and hybrid passivation structures, respectively.

Figure 5a shows the electric field distribution in the channel layer under a drain voltage of 200 V. In comparison with the basic Si_3N_4 passivation structure, the HfO_2 and hybrid passivation structures demonstrated that the peak electric field in the channel layer was reduced and dispersed due to the high dielectric constant of HfO_2. As the peak electric field increased, impact ionization, which causes the generation of electron-hole pairs, became severe. Thus, the redistribution of the electric field effectively improved the V_{BD}. Specifically, the V_{BD} values of the Si_3N_4, HfO_2, and hybrid passivation structures were 232.47, 276.27, and 268.41 V, respectively, as shown in Figure 5b. After applying a voltage of −7 V to the gate to completely turn off the device, the drain voltage when the drain current exceeded 1 mA/mm was defined as the V_{BD}. Figure 5c compares the drain leakage current for the three different passivation structures. Particularly, the structures where HfO_2 is applied to the passivation layer can show that the 2-DEG confinement in the channel region can be improved due to the wide bandgap energy of HfO_2, reducing the leakage current. Therefore, the HfO_2 passivation structure exhibited the least drain leakage current among the three [31,32].

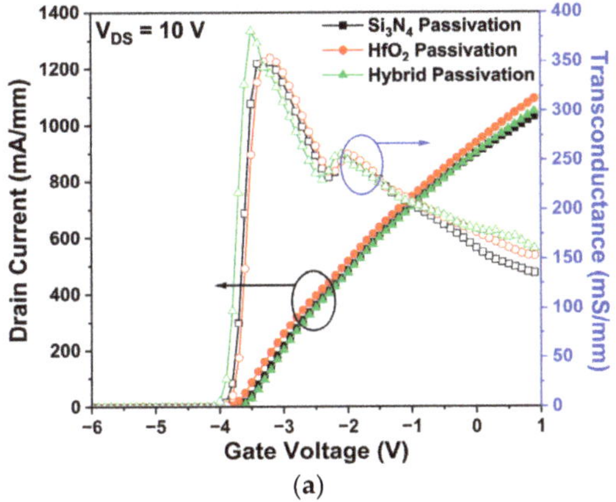

(a)

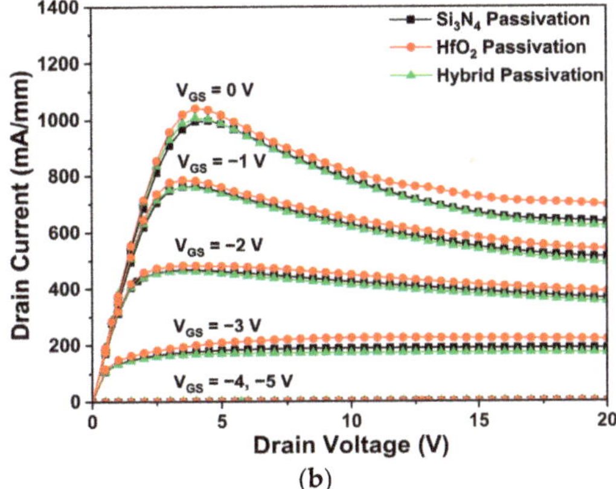

(b)

Figure 4. The DC simulation results of Si_3N_4, HfO_2, and hybrid passivation structures: (**a**) I_{DS}–V_{GS} transfer at V_{DS} = 10 V. The black and blue arrows represent drain current and transconductance, respectively; (**b**) drain current-drain voltage (I_{DS} -V_{DS}) characteristics at V_{GS} = −5, −4, −3, −2, −1, and 0 V.

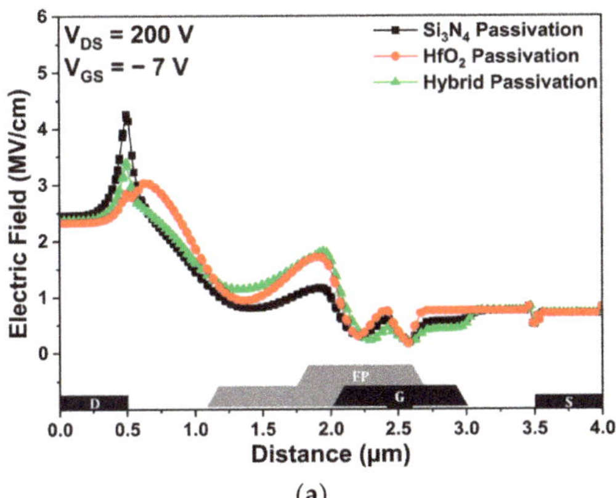

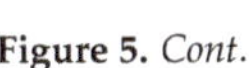
(a)

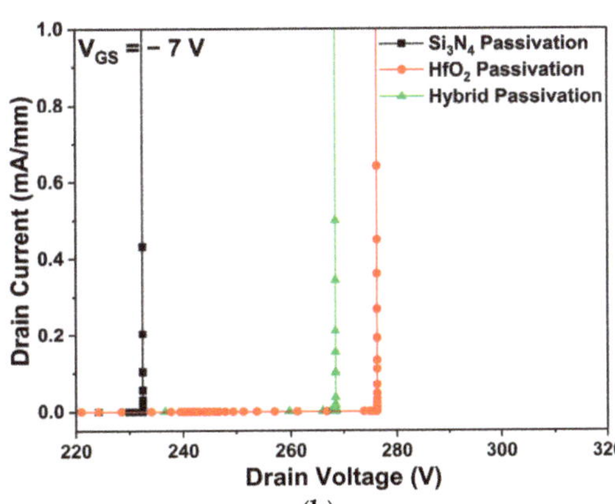

(b)

Figure 5. *Cont.*

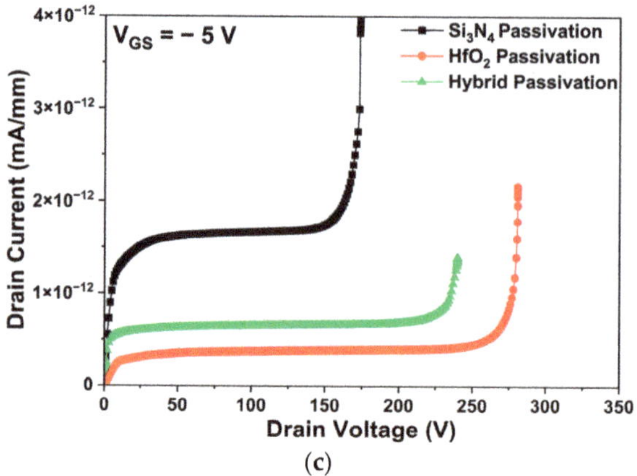

(**c**)

Figure 5. The DC simulation results of Si_3N_4, HfO_2, and hybrid passivation structures: (**a**) electric field distribution in the channel region; (**b**) off-state breakdown voltage; and (**c**) off-state drain leakage current.

3.2.2. Analysis of the RF Characteristics

Figure 6 shows the parasitic capacitance characteristics for Si_3N_4, HfO_2, and hybrid passivation structures. Specifically, the C_{gs} and C_{gd} were obtained at a drain voltage of 20 V and a gate voltage of −2.6 V. As shown in Figure 6a,b, the HfO_2 passivation structure shows the highest C_{gs} and C_{gd}, since the dielectric constant of HfO_2 is larger than that of Si_3N_4, which is explained by Equation (3). In addition, the parasitic capacitance values of the hybrid passivation structure were smaller than that of the HfO_2 passivation structure. This is because the HfO_2 passivation thickness was thinner in the hybrid passivation structure compared to the HfO_2 passivation structure. Therefore, the parasitic capacitances tended to increase as more HfO_2 was used in the passivation layer.

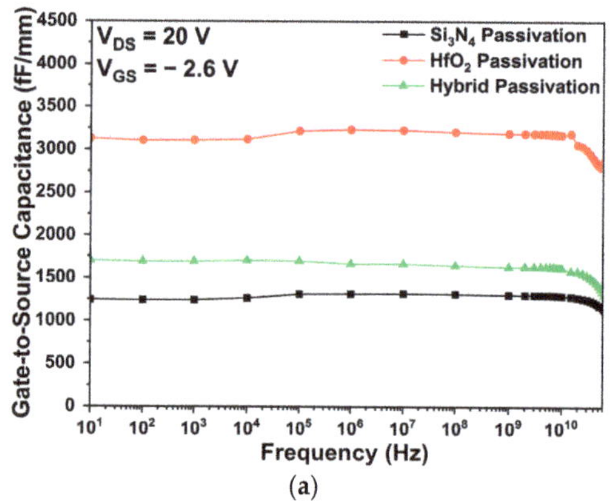

(**a**)

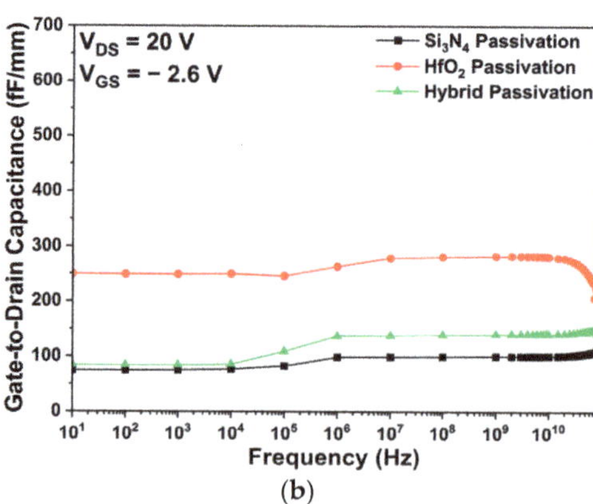

(**b**)

Figure 6. The parasitic capacitance characteristics of Si_3N_4, HfO_2, and hybrid passivation structures: (**a**) gate-to-source capacitance; and (**b**) gate-to-drain capacitance.

Figure 7 represents the simulated f_T and f_{max} of the three different passivation structures. Similarly, as the capacitance simulations, f_T and f_{max}, were obtained at a drain voltage of 20 V and a gate voltage of −2.6 V. More precisely, the f_T values are 24.64, 10.17, and 20.50 GHz for the basic Si_3N_4 passivation, HfO_2 passivation, and hybrid passivation structures, respectively. The f_T values of the HfO_2 and hybrid passivation structures were decreased by 58.7% and 16.8% compared to the basic Si_3N_4 passivation structure, respectively. According to Equation (4), the f_T values of the three passivation structures may have been influenced by the g_m and C_{gs}. In addition, the f_{max} values of the basic Si_3N_4 passivation, HfO_2 passivation, and hybrid passivation structures are 110.28, 48.72, and 88.53 GHz, respectively. It can be seen that f_{max} value of HfO_2 passivation structure significantly decreased as f_T decreased according to Equation (5). Particularly, the f_{max}, which is obtained from the extension line with a slope of −20 dB/decade at the intersection of the maximum stable/available gain (MSG/MAG), becomes 0 dB [33,34].

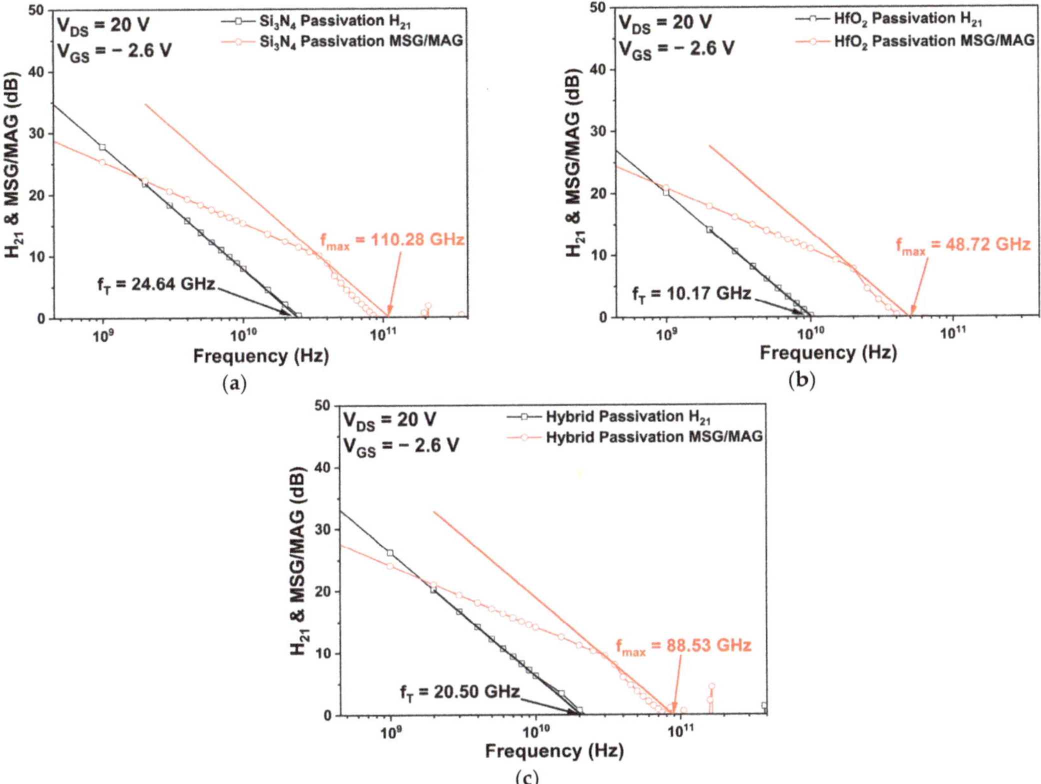

Figure 7. The cut-off frequency (f_T) and maximum oscillation frequency (f_{max}) for different passivation structures: (**a**) Si_3N_4 passivation structure; (**b**) HfO_2 passivation structure; and (**c**) hybrid passivation structure.

Interestingly, these results clearly show that the ratio of HfO_2 in passivation is important for DC and RF performances. As the ratio of HfO_2 increases, the DC characteristics are improved, but the RF characteristics, such as parasitic capacitances and frequency characteristics, are degraded due to the material properties of HfO_2. To improve both DC and RF characteristics, we selected the hybrid passivation structure and then simulated four different 2nd Si_3N_4 passivation thicknesses, i.e., 150, 250, 350, and 450 nm, which will be discussed in Section 3.3. More precisely, to optimize the second Si_3N_4 passivation thickness and calculate the figure-of-merit, we analyzed the operational characteristics including V_{BD}, parasitic capacitances, and frequency characteristics.

3.3. Determination of the Optimum Second Passivation Thickness for Hybrid Structure

3.3.1. Analysis of the DC Characteristics

Figure 8a shows the electric field distribution in the channel region at a drain voltage of 200 V and a gate voltage of −7 V. The peak electric field was not significantly affected by the second passivation thickness. Additionally, the overall electric field distribution also showed no significant difference. Therefore, the V_{BD} values of the various second passivation thickness structures were not changed significantly. As shown in Figure 8b, the V_{BD} was simulated to be 262.00, 268.41, 267.57, and 262.30 V for the hybrid passivation structure with second passivation thicknesses of 150, 250, 350, and 450 nm, respectively. As the field-plate gradually deviates from the channel region, the electric field in the channel cannot be dispersed, resulting in the decrease of V_{BD}. Meanwhile, as the passivation thickness increases, it is expected that V_{BD} would increase, because the passivation can prevent the electric field in the channel region spread by the high electric field adjacent to the gate electrode. For these two reasons, the V_{BD} were slightly enhanced in the second passivation thicknesses of 250 and 350 nm, compared with other structures.

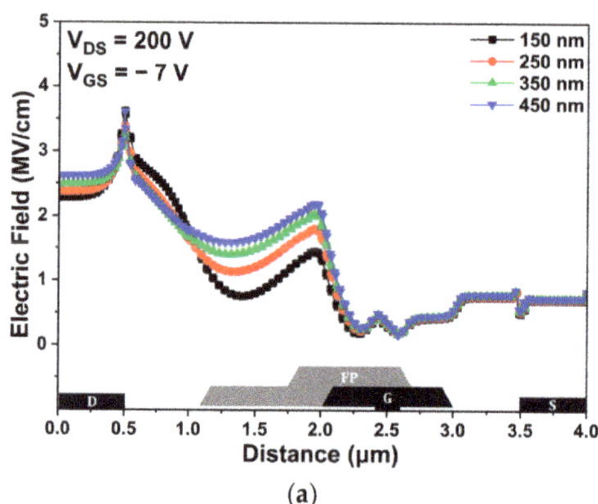

(a)

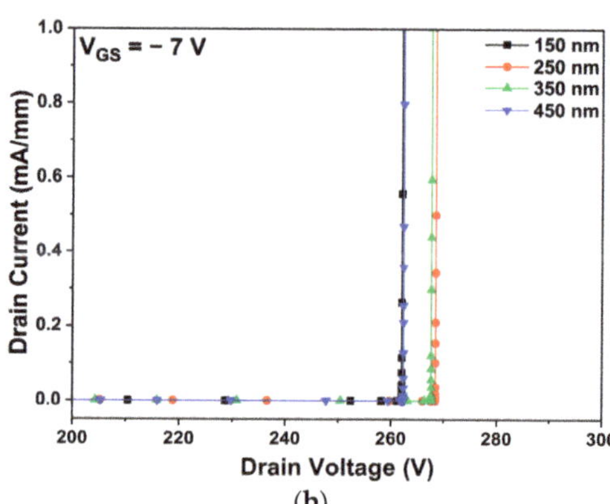

(b)

Figure 8. The DC simulation results of hybrid passivation structure with various second passivation thicknesses: (**a**) electric field distribution in the channel region; and (**b**) off-state breakdown voltage.

3.3.2. Analysis of the RF Characteristics

Figure 9 shows the C_{gs} and C_{gd} of the hybrid passivation structure with various second passivation thicknesses, at a drain voltage of 20 V and a gate voltage of −2.6 V. The second passivation thickness affected the parasitic capacitance values. Specifically, the 150-nm-thick second passivation structure showed the largest C_{gs}, due to the decrease in distance between the gate and source, as shown in Figure 9a. According to Equation (3), as the distance among the electrodes increased, the parasitic capacitances decreased. Therefore, compared to C_{gs}, there is no significant change in C_{gd}, because the gate-to-source distance is much shorter than the gate-to-drain distance. In addition, the 450-nm-thick second passivation structure exhibited a slightly larger C_{gd} than did the other structures, as shown in Figure 9b. The change in materials from air to Si_3N_4 led to an increase in C_{gd} due to dielectric constant of the materials, which is explained by Equation (3).

Figure 10 shows the simulated f_T and f_{max} values for the different second passivation thicknesses at a drain voltage of 20 V and a gate voltage of −2.6 V. When the second passivation thicknesses were 150, 250, 350, and 450 nm, the f_T values in the simulations were 17.92, 20.50, 22.64, and 24.97 GHz, respectively. A decrease in the C_{gs} due to a change in the second passivation thickness led to an increase in f_T, according to Equation (4). Therefore, f_T tended to increase by about 14.4~39.3% as the second passivation thickness was extended by each 100-nm-step. The f_{max} values were simulated to be 78.50, 88.53, 91.47, and 106.39 GHz for the hybrid passivation structure with the second passivation thicknesses of 150, 250, 350, and 450 nm, respectively. Comparing the f_{max} values of the hybrid passivation structures based on the different second passivation thicknesses, it can be demonstrated that the f_{max} values increased by 12.8~35.5% with each 100-nm-step increase in the second passivation thickness. According to Equation (5), the f_{max} values were mainly influenced by the increase in f_T because there was no significant change in C_{gd}. Throughout these results, we confirmed the dependence of frequency characteristics in relation to the second passivation thickness.

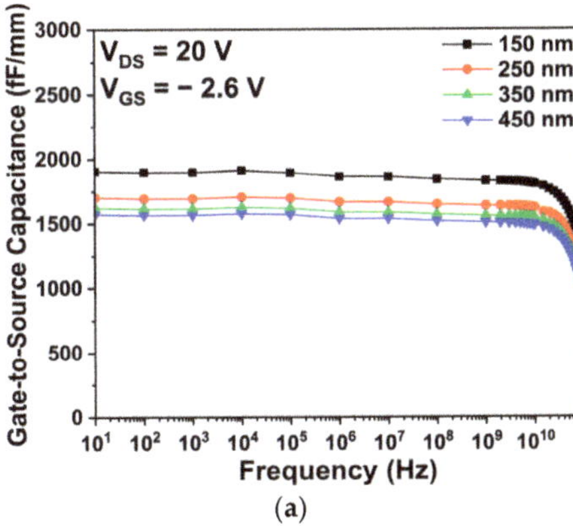

(a)

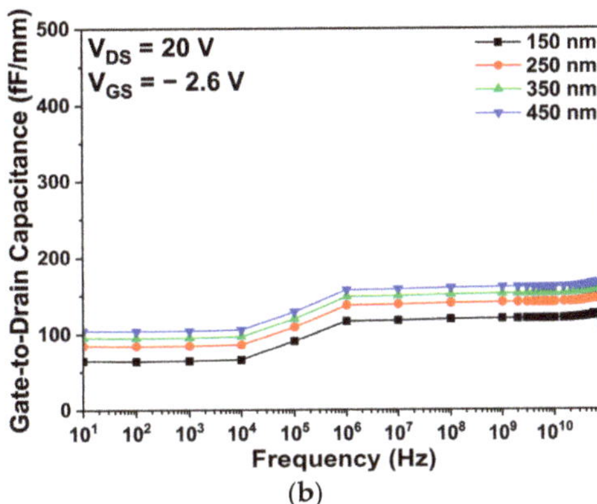

(b)

Figure 9. The parasitic capacitance characteristics of the hybrid passivation structure with various second passivation thicknesses: (**a**) gate-to-source capacitance; and (**b**) gate-to-drain capacitance.

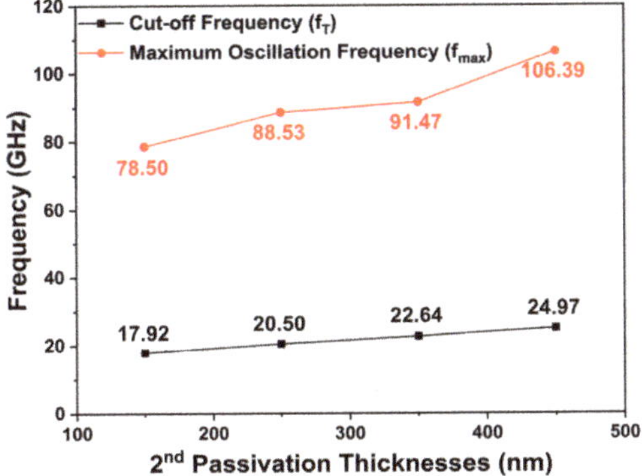

Figure 10. The simulated f_T and f_{max} as a function of the second passivation thicknesses at V_{DS} = 20 V and V_{GS} = −2.6 V.

4. Discussion

In this article, we simulated the DC and RF characteristics of various passivation structures. Additionally, we analyzed the hybrid passivation structure by changing the second passivation thickness. Based on these results, we first calculated the LFOM and JFOM to investigate the performance of the device for the various passivation structures. Table 4 provides a summary of the DC and RF characteristics, including the figure-of-merit for the four different passivation structures. More precisely, the LFOM and JFOM of the basic Si_3N_4 passivation structures were 13.44 MW/mm and 5.73 THz-V, respectively. The HfO_2 passivation structure increased the LFOM by 48% and decreased the JFOM by 39% compared with the basic Si_3N_4 passivation structure. In comparison with the basic Si_3N_4

passivation structure, analysis of the hybrid passivation structure showed that the LFOM was increased by up to 35% and the JFOM was decreased by up to 4%.

Table 4. A summary of the DC and RF characteristics of various passivation structure HEMTs.

Parameters	Units	Si_3N_4	HfO_2	Hybrid	
First/second passivation thickness	nm	50/250	50/250	50/250	50/350
On-resistance (R_{on})	Ω-mm	4.02	3.84	3.97	4.16
Breakdown voltage (V_{BD})	V	232.47	276.27	268.41	267.57
Cut-off frequency (f_T)	GHz	24.64	10.17	20.50	22.64
Maximum oscillation frequency (f_{max})	GHz	110.27	48.72	88.53	91.47
Standard lateral figure-of-merit (LFOM)	MW/mm	13.44	19.93	18.15	17.21
Johnson's figure-of-merit (JFOM)	THz-V	5.73	2.81	5.50	6.06

Subsequently, the LFOM values for the hybrid passivation structure of different second passivation thicknesses were estimated to be 17.93, 18.15, 17.68, and 15.53 MW/mm, respectively. In addition, except for the hybrid passivation structure with 450-nm-thick second Si_3N_4 passivation, the LFOM values of the other hybrid passivation structures had improved by more than 28%, compared to the basic Si_3N_4 passivation structure. Further, we measured the JFOM values for the hybrid passivation structures of different second passivation thicknesses, which were 4.70, 5.50, 6.06, and 6.55 THz-V, respectively. As the second passivation thickness increased, the JFOM values also increased.

5. Conclusions

In this study, using TCAD simulation, we analyzed the operational characteristics of AlGaN/GaN HEMTs in accordance with changes of passivation materials and thicknesses. Before analyzing the various passivation structures, all the simulation and material parameters were precisely set through mapping with the measurement data of the fabricated device to ensure the reliability of the simulated data. Based on the simulation results, we suggest an optimized hybrid structure of HEMT which adopts a 50-nm-thick first HfO_2 passivation and a 350-nm-thick second Si_3N_4 passivation. Unlike other general structures such as the field-plate in the HEMT, we confirmed that the hybrid passivation structure of the HEMT with suitable passivation thickness could enhance both the DC and RF performances, including the LFOM and JFOM. Consequently, the simulation results clearly show that HfO_2 as a passivation material with a second passivation thickness suitable for the AlGaN/GaN HEMTs can be a promising candidate for future high-power and high-frequency applications.

Author Contributions: Conceptualization and writing—original draft preparation, J.-H.C.; software and investigation, W.-S.K.; formal analysis and data curation, D.K.; validation and formal analysis, J.-H.K.; formal analysis and investigation, J.-H.L.; validation and investigation, K.-Y.K.; validation and formal analysis, B.-G.M.; resources and investigation, D.M.K.; supervision, funding acquisition, resources, and writing—review and editing, H.-S.K. All authors have read and agreed to the published version of the manuscript.

Funding: This work was partly supported by an Institute of Information & Communications Technology Planning & Evaluation (IITP) grant funded by the government of the Republic of Korea (MSIT) under grant No. 2021-0-00760, as well as the Institute of Civil Military Technology Cooperation, funded by the Defense Acquisition Program Administration and the Ministry of Trade, Industry and Energy of the government of the Republic of Korea, under grant No. 22-CM-15.

Conflicts of Interest: The authors declare no conflict of interest.

References

1. Zhang, N.Q.; Keller, S.; Parish, G.; Heikman, S.; DenBaars, S.P.; Mishra, U.K. High breakdown GaN HEMT with overlapping gate structure. *IEEE Electron Device Lett.* **2000**, *21*, 421–423. [CrossRef]
2. Meneghini, M.; Rossetto, I.; Santi, C.D.; Rampazzo, F.; Tajalli, A.; Barbato, A.; Ruzzarin, M.; Borga, M.; Canato, E.; Zanoni, E.; et al. Reliability and failure analysis in power GaN-HEMTs: An overview. In Proceedings of the IEEE International Reliability Physics Symposium (IRPS), Monterey, CA, USA, 2–6 April 2017.
3. Mishra, U.K.; Shen, L.; Kazior, T.E.; Wu, Y.F. AlGaN/GaN HEMTs-an overview of device operation and applications. *Proc. IEEE* **2002**, *90*, 1022–1031. [CrossRef]
4. Del Alamo, J.A.; Joh, J. GaN HEMT reliability. *Microelectron. Reliab.* **2009**, *49*, 1200–1206. [CrossRef]
5. Kim, J.G.; Cho, C.; Kim, E.; Hwang, J.S.; Park, K.H.; Lee, J.H. High Breakdown Voltage and Low-Current Dispersion in AlGaN/GaN HEMTs with High-Quality AlN Buffer Layer. *IEEE Trans. Electron Devices* **2021**, *68*, 1513–1517. [CrossRef]
6. Sodan, V.; Oprins, H.; Stoffels, S.; Baelmans, M.; Wolf, I.D. Influence of Field-Plate Configuration on Power Dissipation and Temperature Profiles in AlGaN/GaN on Silicon HEMTs. *IEEE Trans. Electron Devices* **2015**, *62*, 2416–2422. [CrossRef]
7. Saito, W.; Takada, Y.; Kuraguchi, M.; Tsuda, K.; Omura, I.; Omura, T. High breakdown voltage AlGaN-GaN Power-HEMT design and high current density switching behavior. *IEEE Trans. Electron Devices* **2003**, *50*, 2528–2531. [CrossRef]
8. Shi, N.; Wang, K.; Zhou, B.; Weng, J.; Cheng, Z. Optimization AlGaN/GaN HEMT with Field Plate Structures. *Micromachines* **2022**, *13*, 702. [CrossRef]
9. Saadat, O.I.; Chung, J.W.; Piner, E.L.; Palacios, T. Gate-First AlGaN/GaN HEMT Technology for High-Frequency Applications. *IEEE Electron Device Lett.* **2009**, *30*, 1254–1256. [CrossRef]
10. Ahsan, S.A.; Ghosh, S.; Sharma, K.; Dasgupta, A.; Khandelwal, S.; Chauhan, Y.S. Capacitance modeling in dual field-plate power GaN HEMT for accurate switching behavior. *IEEE Trans. Electron Devices* **2015**, *63*, 565–572. [CrossRef]
11. Kwak, H.T.; Chang, S.B.; Kim, H.J.; Jang, K.W.; Yoon, H.S.; Lee, S.H.; Lim, J.W.; Kim, H.S. Operational Improvement of AlGaN/GaN High Electron Mobility Transistor by an Inner Field-Plate Structure. *Appl. Sci.* **2018**, *8*, 974. [CrossRef]
12. Sohel, S.H.; Xie, A.; Beam, E.; Xue, H.; Razzak, T.; Bajaj, S.; Cao, Y.; Lee, C.; Lu, W.; Rajan, S. Polarization Engineering of AlGaN/GaN HEMT With Graded InGaN Sub-Channel for High-Linearity X-Band Applications. *IEEE Electron Device Lett.* **2019**, *40*, 522–525. [CrossRef]
13. Liu, X.; Chiu, H.C.; Liu, C.H.; Kao, H.L.; Chiu, C.W.; Wang, H.C.; Ben, J.; He, W.; Huang, C.R. Normally-off p-GaN Gated AlGaN/GaN HEMTs Using Plasma Oxidation Technique in Access Region. *IEEE J. Electron Devices Soc.* **2020**, *8*, 229–234. [CrossRef]
14. Kuzmik, J.; Pozzovivo, G.; Abermann, S.; Carlin, J.; Gonschorek, M.; Feltin, E.; Grandjean, N.; Bertagnolli, E.; Strasser, G.; Pogany, D. Technology and Performance of InAlN/AlN/GaN HEMTs with Gate Insulation and Current Collapse Suppression Using ZrO_2 or HfO_2. *IEEE Trans. Electron Devices* **2008**, *55*, 937–941. [CrossRef]
15. Coltrin, M.E.; Baca, A.G.; Kaplar, R.J. Analysis of 2D transport and performance characteristics for lateral power devices based on AlGaN alloys. *ECS J. Solid State Sci. Technol.* **2017**, *6*, S3114–S3118. [CrossRef]
16. Downey, B.P.; Meyer, D.J.; Katzer, D.S.; Roussos, J.A.; Pan, M.; Gao, X. SiN_x/InAlN/AlN/GaN MIS-HEMTs with 10.8THzV Johnson Figure of Merit. *IEEE Electron Device Lett.* **2014**, *35*, 527–529. [CrossRef]
17. Sanyal, I.; Lin, E.; Wan, Y.; Chen, K.; Tu, P.; Yeh, P.; Chyi, J. AlInGaN/GaN HEMTs with High Johnson's Figure-of-Merit on Low Resistivity Silicon Substrate. *IEEE J. Electron Devices Soc.* **2020**, *9*, 130–136. [CrossRef]
18. Yoon, H.S.; Min, B.G.; Lee, J.M.; Kang, D.M.; Ahn, H.K.; Kim, H.; Lim, J. Microwave Low-Noise Performance of 0.17 μm Gate-Length AlGaN/GaN HEMTs on SiC with Wide Head Double-Deck T-Shaped Gate. *IEEE Electron Device Lett.* **2016**, *37*, 1407–1410. [CrossRef]
19. Silvaco. Material Dependent Physical Models. In *Atlas User's Manual Device Simulation Software*; Silvaco Inc.: Santa Clara, CA, USA, 2016; pp. 519–523.
20. Kim, H.J.; Jang, K.W.; Kim, H.S. Operational Characteristics of Various AlGaN/GaN High Electron Mobility Transistor Structures Concerning Self-Heating Effect. *J. Nanosci. Nanotechnol.* **2019**, *19*, 6016–6022. [CrossRef]
21. Raja, P.V.; Subramani, N.K.; Gaillard, F.; Bouslama, M.; Sommet, R.; Nallatamby, J. Identification of Buffer and Surface Traps in Fe-Doped AlGaN/GaN HEMTs Using Y21 Frequency Dispersion Properties. *Electronics* **2021**, *10*, 3096. [CrossRef]
22. Vitanov, S.; Palankovski, V.; Maroldt, S.; Quay, R. High-temperature modeling of algan/gan hemts. *Solid-State Electron.* **2010**, *54*, 1105–1112. [CrossRef]
23. Lee, J.H.; Choi, J.H.; Kang, W.S.; Kim, D.; Min, B.G.; Kang, D.M.; Choi, J.H.; Kim, H.S. Analysis of Operational Characteristics of AlGaN/GaN HEMT with Various Slant-Gate-Based Structures: A Simulation Study. *Micromachines* **2022**, *13*, 1957. [CrossRef] [PubMed]
24. Belkacemi, K.; Hocine, R. Efficient 3D-TLM modeling and simulation for the thermal management of microwave AlGaN/GaN HEMT used in high power amplifiers SSPA. *J. Low Power Electron Appl.* **2018**, *8*, 23. [CrossRef]
25. Brady, R.G.; Oxley, C.H.; Brazil, T.J. An Improved Small-Signal Parameter-Extraction Algorithm for GaN HEMT Devices. *IEEE Trans. Microw. Theory Tech.* **2008**, *56*, 1535–1544. [CrossRef]
26. Yoon, H.S.; Min, B.G.; Lee, J.M.; Kang, D.M.; Ahn, H.K.; Kim, H.C.; Lim, J.W. Wide head T-shaped gate process for low-noise AlGaN/GaN HEMTs. In Proceedings of the CS MANTECH Conference, Scottsdale, AZ, USA, 18–21 May 2015.

27. Saito, W.; Nitta, T.; Kakiuchi, Y.; Saito, Y.; Tusda, K.; Omura, I.; Yamaguchi, M. On-Resistance Modulation of High Voltage GaN HEMT on Sapphire Substrate Under High Applied Voltage. *IEEE Electron Device Lett.* **2007**, *28*, 676–678. [CrossRef]
28. Nuttinck, S.; Gebara, E.; Laskar, J.; Harris, H.M. Study of self-heating effects, temperature-dependent modeling, and pulsed load-pull measurements on GaN HEMTs. *IEEE Trans. Microw. Theory Tech.* **2001**, *49*, 2413–2420. [CrossRef]
29. Cioni, M.; Zagni, N.; Selmi, L.; Meneghesso, G.; Meneghini, M.; Zanoni, E.; Chini, A. Electric Field and Self-Heating Effects on the Emission Time of Iron Traps in GaN HEMTs. *IEEE Trans. Electron Devices* **2021**, *68*, 3325–3332. [CrossRef]
30. Jang, K.W.; Hwang, I.T.; Kim, H.J.; Lee, S.H.; Lim, J.W.; Kim, H.S. Thermal analysis and operational characteristics of an AlGaN/GaN High electron mobility transistor with copper-filled structures: A simulation study. *Micromachines* **2019**, *11*, 53. [CrossRef]
31. Kabemura, T.; Ueda, S.; Kawada, Y.; Horio, K. Enhancement of Breakdown Voltage in AlGaN/GaN HEMTs: Field Plate Plus High-*k* Passivation Layer and High Acceptor Density in Buffer Layer. *IEEE Trans. Electron Devices* **2018**, *65*, 3848–3854. [CrossRef]
32. Hanawa, H.; Onodera, H.; Nakajima, A.; Horio, K. Numerical Analysis of Breakdown Voltage Enhancement in AlGaN/GaN HEMTs with a High-k Passivation Layer. *IEEE Trans. Electron Devices* **2014**, *61*, 769–775. [CrossRef]
33. Cheng, J.; Rahman, M.W.; Xie, A.; Xue, H.; Sohel, S.H.; Beam, E.; Lee, C.; Yang, H.; Wang, C.; Cao, Y.; et al. Breakdown Voltage Enhancement in ScAlN/GaN High-Electron-Mobility Transistors by High-*k* Bismuth Zinc Niobate Oxide. *IEEE Trans. Electron Devices* **2021**, *68*, 3333–3338. [CrossRef]
34. Chung, J.W.; Hoke, W.E.; Chumbes, E.M.; Palacios, T. AlGaN/GaN HEMT with 300-GHz f_{max}. *IEEE Electron Device Lett.* **2010**, *31*, 195–197. [CrossRef]

micromachines

Article

Analysis of Noise-Detection Characteristics of Electric Field Coupling in Quartz Flexible Accelerometer

Zhigang Zhang [1,2,3], Dongxue Zhao [1,2,3], Huiyong He [1,2,3], Lijun Tang [1,2,3,*] and Qian He [1,2,3]

1 School of Physics and Electronic Science, Changsha University of Science and Technology, Changsha 410114, China
2 Hunan Provincial Key Laboratory of Flexible Electronic Materials Genome Engineering, Changsha 410114, China
3 Key Laboratory of Electromagnetic Environment Monitoring and Modeling in Hunan Province, Changsha 410114, China
* Correspondence: tanglj@csust.edu.cn; Tel.: +86-13974833567

Abstract: The internal electric field coupling noise of a quartz flexible accelerometer (QFA) restricts the improvement of the measurement accuracy of the accelerometer. In this paper, the internal electric field coupling mechanism of a QFA is studied, an electric field coupling detection noise model of the accelerometer is established, the distributed capacitance among the components of the QFA is simulated, the structure of the detection noise transfer system of different carrier modulation differential capacitance detection circuits is analyzed, and the influence of each transfer chain on the detection noise is discussed. The simulation results of electric field coupling detection noise show that the average value of detection noise can reach 41.7 μg, which is close to the effective resolution of the QFA, 50 μg. This confirms that electric field coupling detection noise is a non-negligible factor affecting the measurement accuracy of the accelerometer. A method of adding a high-pass filter to the front of the phase-shifting circuit is presented to suppress the noise of electric field coupling detection. This method attenuates the average value of the detected noise by about 78 dB, and reduces the average value of the detected noise to less than 0.1 μg, which provides a new approach and direction for effectively breaking through the performance of the QFA.

Keywords: quartz flexible accelerometer; electric field coupling noise; lock-in amplifier circuit

Citation: Zhang, Z.; Zhao, D.; He, H.; Tang, L.; He, Q. Analysis of Noise-Detection Characteristics of Electric Field Coupling in Quartz Flexible Accelerometer. *Micromachines* **2023**, *14*, 535. https://doi.org/10.3390/mi14030535

Academic Editor: Ha Duong Ngo

Received: 1 February 2023
Revised: 22 February 2023
Accepted: 22 February 2023
Published: 25 February 2023

1. Introduction

The accuracy of quartz flexible accelerometers is an important factor that affects the navigation performance of inertial navigation systems. According to public data, high-precision quartz flexible accelerometers such as QA3000-30 [1], AI-Q-2010 [2], and A600 [3] have been widely used in inertial navigation systems. The performances of these three QFAs are shown in Table 1. Among them, the QA3000-30 accelerometer developed by Honeywell in the 1990s has an accuracy of 1 μg, making it the most accurate and reliable QFA currently available. However, after more than 30 years of development, the effective accuracy of the QFA remains around 1 μg without a significant breakthrough [4]. The stagnation of the performance development of the QFA has gradually become the obstacle to overcome in improving the navigation performance of inertial navigation systems.

The noise in a QFA system mainly includes the noise caused by the change in external temperature and other environmental parameters, the mechanical thermal noise, the resonance noise [5], and the coupling noise [6] caused by the operation of the differential capacitance detection circuit and torquer drive circuit in the system. The impact of this noise on the performance of the accelerometer cannot be ignored. In recent years, there have been many studies on the effect of noise caused by temperature change on the performance of accelerometers [7–11]. The newly researched temperature compensation method maintains the stability of the cold start output at about 30 μg [12], and increases the stability of 1 g acceleration output to 14.6 μg [13], which significantly improves the temperature stability of the QFA.

Table 1. Main performance characteristics of QA3000-300, AI-Q-2010 and A600.

	QA3000-30	AI-Q-2010	A600
Input Range (g)	±60	±60	±60
Bias (mg)	4	4	4
Bias repeatability (μg)	40	550	20
Bias temperature sensitivity (μg/°C)	15	30	30
Scale Factor (mA/g)	1.20~1.46	1.20~1.46	1.20~1.46
Threshold (μg)	1	1	1
Bandwidth (Hz)	300	300	1000
Intrinsic Noise (μg RMS)	1500	1500	1500

In order to improve the accuracy of the QFA, the internal noise of the QFA system has attracted much attention from researchers, and a lot of meaningful work has been carried out on the differential capacitance detection circuit [14,15], the analysis of the noise characteristics of the drive circuit [16,17], and the design and optimization of the circuit structure. Huang et al. established a noise model of the readout circuit, analyzed the noise of the charge-discharge detection circuit and its equivalent acceleration formula, and found that the noise of the charge-discharge method is about 1 μg [18]. Ran et al. proposed a digital closed-loop QFA signal-detection method based on the AC bridge, and Chen et al. designed a differential capacitance detection method based on the capacitor bridge [19]. Both methods achieved a resolution for the detection circuit of 0.0018 pF, which matches the measurement of 1 μg resolution acceleration [20]. In these studies, the researchers only focused on the noise optimization of a single detection circuit or driving circuit, and did not pay attention to the coupling relationship between them. When a QFA is working, the driving signal in the torquer coil changes rapidly, and the electromagnetic radiation generated is coupled to the differential capacitance plate, which is introduced into the differential capacitance detection circuit to cause interference. The existence of electromagnetic coupling noise affects the performance of the QFA [21]. In this research, we studied the internal electric field coupling mechanism of a QFA, established an accelerometer electric field coupling detection noise model, analyzed the characteristics of the electric field coupling detection noise, and explored the influencing factors and suppression methods of the electric field coupling detection noise.

2. Analysis of Electric Field Coupling in QFA

2.1. Mechanism of Electric Field Coupling in QFA

A QFA is mainly composed of a quartz pendulum reed component, differential capacitance sensor, torquer, shell, etc. [22], and its structure is shown in Figure 1a. The quartz pendulum assembly, differential capacitance sensor, torquer, and drive circuit constitute the closed-loop system shown in Figure 1b [23]. The transfer function $H(s)$ of the closed-loop system can be expressed as:

$$H(s) = \frac{I_T}{a_i} = \frac{K_B\theta(s)K_sK_aM(s)K_I}{1+\theta(s)K_sK_aM(s)K_IK_T} \quad (1)$$

where K_B is the pendulum property of the pendulum reed component, and $\theta(s)$, K_s, K_a, $M(s)$, K_I, and K_T are the transfer functions of the pendulum reed component, differential capacitance sensor, differential capacitance detection circuit, torquer control system, torquer driving circuit, and torquer, respectively. The pendulum reed transfer function $\theta(s)$ of QFA is:

$$\theta(s) = \frac{1}{JS^2+CS+K} \quad (2)$$

where J is the rotational inertia, C is the damping factor, and K is the elastic factor. In the closed-loop system, the input acceleration a_i can be obtained by measuring the drive current I_T, and its direction depends on the drive-current direction.

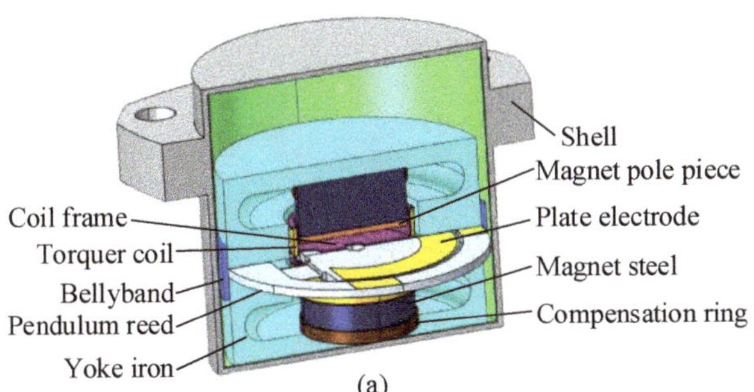

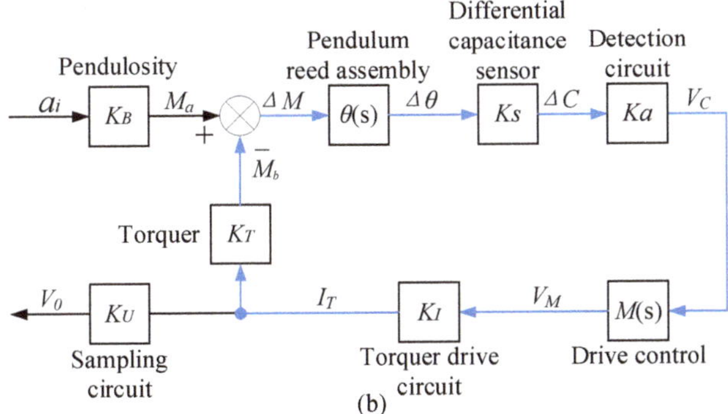

Figure 1. (**a**) Structure of QFA; (**b**) Closed-loop system structure of QFA.

In the QFA, the shell, yoke iron, torquer coil, differential capacitor plate, and other components are made of metal materials. These components are independent of each other and do not make contact with each other. When the accelerometer works, there is a potential difference in electricity between different components, which forms the distributed capacitance shown in Figure 2a. C_{d1}, C_{d2}, and C_{d3} are the distributed capacitances between the torquer coil and the plates of the differential capacitor; C_{bb} is the distributed capacitance between the top and bottom plates of the differential capacitor; C_{e1}, C_{e2}, and C_{e3} are the distributed capacitances between the three plates of differential capacitance and the accelerometer shell; and C_{e4} is the distributed capacitance between the accelerometer shell and the torquer coil.

The distributed capacitance couples the torquer drive current I_T to the differential capacitance detection process to form the detection noise V_n. Based on this, the equivalent circuit model of electric field coupling detection noise inside the accelerometer is established, as shown in Figure 2b. The controlled current source I_T is the current output of the torquer drive circuit; L_T and R_T are the resistance and inductance of the torquer coil, respectively; and R_{CS} is the current sampling resistance.

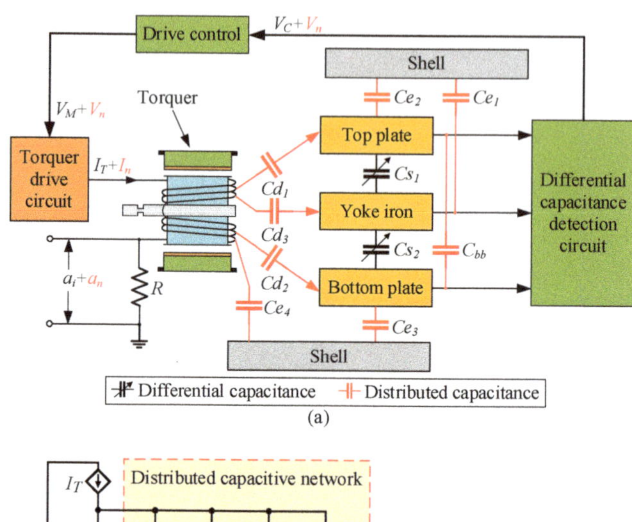

(a)

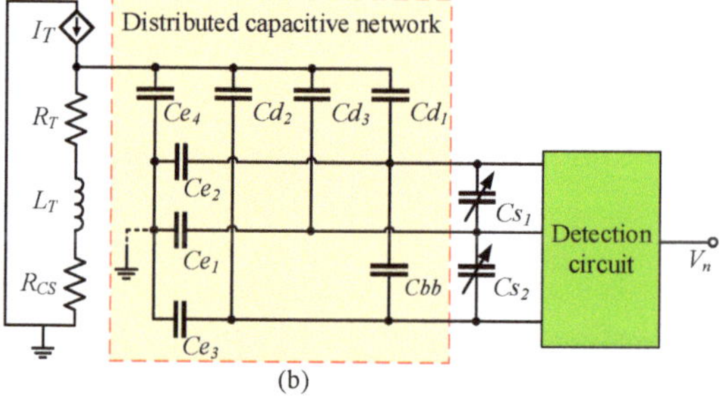

(b)

Figure 2. (**a**) Schematic diagram of distributed capacitance among various components of QFA; (**b**) Equivalent circuit model of electric field coupling detection noise in QFA.

2.2. Simulation of Distributed Capacitance in QFA

When analyzing the noise characteristics of accelerometer electric field coupling detection, it is necessary to determine the value and variation law of each distributed capacitance. The value of the distributed capacitance in the QFA mainly depends on the directly opposite area and distance of each component. Only the pendulum reed in each component will move with the change in input acceleration, which will change the area and distance between the components. Therefore, it is only necessary to study the relationship between the distributed capacitance and the displacement of the pendulum reed. In this study, the finite-element simulation method was used for analysis and calculation.

The simulation model of the QFA was established according to the assembly structure and materials of a typical QFA. The main performance of the typical QFA is shown in Table 2.

Table 2. Main performance characteristics of typical QFA.

Performance	Value
Input range (g)	±10
Bias (mg)	5
Bias repeatability (μg)	30
Bias temperature sensitivity (μg/°C)	30
Scale factor (mA/g)	1.00~1.20
Threshold (μg)	5
Bandwidth (Hz)	300
Intrinsic noise (μg RMS)	3000

The geometric dimensions of the main structures are shown in Table 3.

Table 3. Main geometric dimensions of QFA.

Size Name	Value (mm)	Size Name	Value (mm)
Shell height	25.00	Pendulum reed thickness	0.72
Shell diameter	38.20	Pendulum reed diameter	17.40
Meter head height	16.22	Coating area of pendulum reed	90.19 (mm^2)
Meter head diameter	23.40	Length of flexible beam	2.80
Torquer coil height	2.40	Width of flexible beam	3.60
Torquer coil diameter	10.60	Thickness of flexible beam	0.02
Distance between upper and lower yoke iron	0.02	Flexible beam spacing	2.50

The material type and relative permittivity of each component in the simulation model are shown in Table 4.

Table 4. Material parameters of components of QFA.

Part Name	Material Type	Relative Permittivity
Shell	1Cr18Ni9Ti	1.00
Pendulum reed	JGS1	3.83
Coating film	Au	1.00
Torquer coil	Cu	1.00
Coil frame	Al2O3	9.50
Magnet steel	XGS240/46	1.00
Magnet pole piece	1J50	1.00
Compensation ring	1J38	1.00
Yoke iron	4J36	1.00
Bellyband	4J36	1.00
Adhesive tape	3M8992	3.10
Underfill	DG-3S	2.70
Filling gas	Air	1.00

Voltage excitation was applied to each component according to the actual working condition of the accelerometer head. The voltage excitation settings are shown in Table 5.

Table 5. Excitation voltage value of each component.

Part Name	Static Pole Plate	Torquer Coil	Top Plate	Bottom Plate	Shell
Voltage (V)	0.00	6.00	5.00	−5.00	0.00

A negative displacement with a step length of 0.5 μm was applied to the quartz pendulum component along the sensitive axis, and the variation law of differential capacitance and distributed capacitance of the pendulum component was analyzed from the balance position to the deviation range of 19.00 μm. The accuracy of the simulation calculation was set to 1%, which considered both simulation efficiency and data accuracy.

The simulation results of differential capacitance and distributed capacitance are shown in Figure 3. The differential capacitance C_{s1} decreases monotonically and C_{s2} increases monotonically with the increase in pendulum reed displacement. When the movement of the pendulum reed is very small, the decrease in C_{s1} is approximately equal to the increase in C_{s2}, and can be approximately expressed as:

$$\begin{cases} C_{s1} = C_0 - \Delta C \\ C_{s2} = C_0 + \Delta C \end{cases} \tag{3}$$

where ΔC is the change value of the differential capacitance, and C_0 is the initial value of the differential capacitance when the pendulum reed is in the balance position, C_0 = 40 pF.

Equation (3) conforms to the characteristics of the QFA differential capacitance sensor [24]. C_{d1}, C_{d2}, and C_{d3} increased or decreased with the increase in pendulum displacement, but the variation was only ±0.05 pF; C_{e1} and C_{e2} have no obvious change trend with the increase in pendulum displacement; C_{e3} and C_{e4} have a decreasing trend, but the change is very small; C_{bb} has a decreasing trend with the increase in pendulum displacement, and the variation is 0.03 pF.

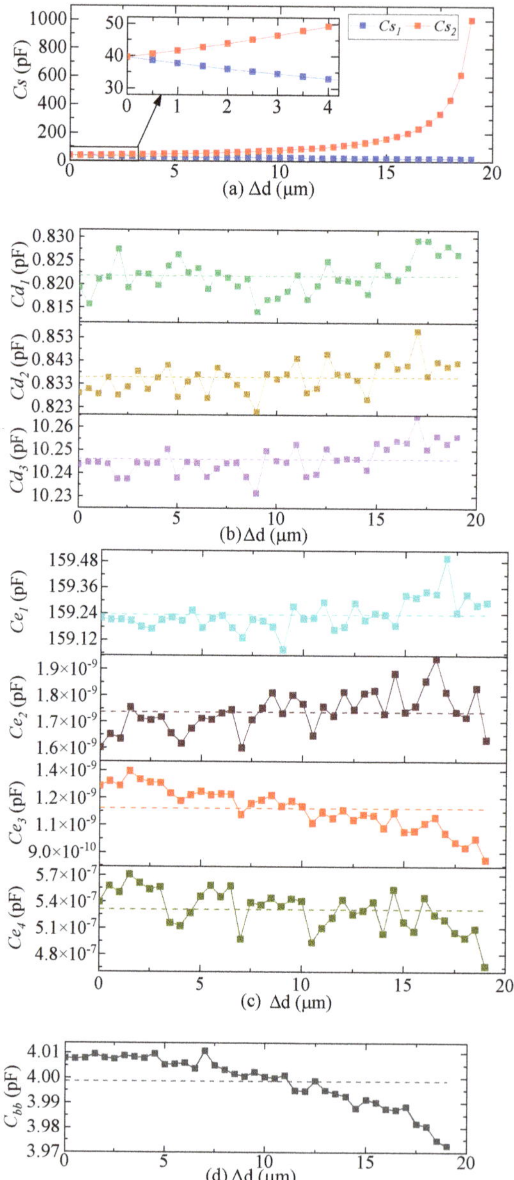

Figure 3. Simulation results of differential capacitance and distributed capacitance. (**a**) Simulation results of differential capacitance C_{s1} and C_{s2}; (**b**) Simulation results of distributed capacitance C_{d1}, C_{d2}, and C_{d3}; (**c**) Simulation results of distributed capacitance C_{e1}, C_{e2}, C_{e3}, and C_{e4}; (**d**) Simulation results of distributed capacitance C_{bb}.

The simulation results show that, within a certain error range, the value of the distributed capacitance is independent of the movement of the pendulum reed, which can be explained by the structure and working principles of the QFA. When the pendulum reed moves, the torquer coil wound on the coil frame will not move relative to the pendulum reed because it is bonded to both sides of it. The distributed capacitance C_{d1} and C_{d2} between the torquer coil and the top and bottom plates of the differential capacitor are independent of the movement of the pendulum reed. When the torquer coil moves with the pendulum, it is always between the upper and lower yoke iron. When the area of one side decreases, the area of the opposite side increases, and the two areas increase and decrease equally. The distributed capacitance C_{d3} between the torquer coil and yoke iron remains unchanged. The metal coating on both sides of the pendulum reed is always static, and the distributed capacitance C_{bb} between the top and bottom plates of the differential capacitor is fixed. The yoke iron and the shell are fixed, and the distributed capacitance C_{e1} is fixed. In addition, the upper and lower yoke iron and the bellyband are welded using a laser to form an almost closed space, which produces an electrostatic shielding effect. The distributed capacitances C_{e2}, C_{e3}, and C_{e4} between the components inside the enclosed space and the shell are very small. Therefore, when the electric field coupling detection noise equivalent circuit model is used to analyze the detection noise, the very small distributed capacitances C_{e2}, C_{e3}, and C_{e4} can be ignored. The distributed capacitance values are shown in Table 6 and can be regarded as fixed values.

Table 6. Distributed capacitance value.

Capacitance Label	C_{d1}	C_{d2}	C_{d3}	C_{bb}	C_{e1}	C_0
Value (pF)	0.8	10.0	0.8	4.0	159.0	40.0

3. Analysis of Detection Noise Transfer System Structure

According to the equivalent circuit model of electric field coupling detection noise, the noise is related to the type of differential capacitance detection circuit. At present, the common differential capacitance detection circuits mainly include capacitance divider type, switch capacitance integral type, ring diode type and carrier modulation type detection circuits. Among them, the carrier modulation detection circuit is widely used in QFAs. The carrier modulation detection circuits are divided into single-channel carrier modulation (SCM) and dual-channel carrier modulation (DCM) detection circuits. In addition, the shell has two connection modes: grounded and float. Shell grounding is also a common electromagnetic shielding method. The float shell is designed to prevent the internal components from being destroyed in use. Different types of differential capacitance detection circuits and the connection mode of the accelerometer shell constitute multiple combinations of noise transfer paths.

3.1. Detection Noise Transfer System Structure of SCM Detection Circuit

When an SCM detection circuit is used to detect differential capacitance, the structure of the detection circuit is as shown in Figure 4a. The single-channel high-frequency sinusoidal carrier signal V_s is input to the differential capacitance fixed plate, and the signals of the two moving plates pass through the charge amplification circuit and the differential amplification circuit to obtain the modulated signal V_a, whose amplitude reflects the variation in the differential capacitance ΔC.

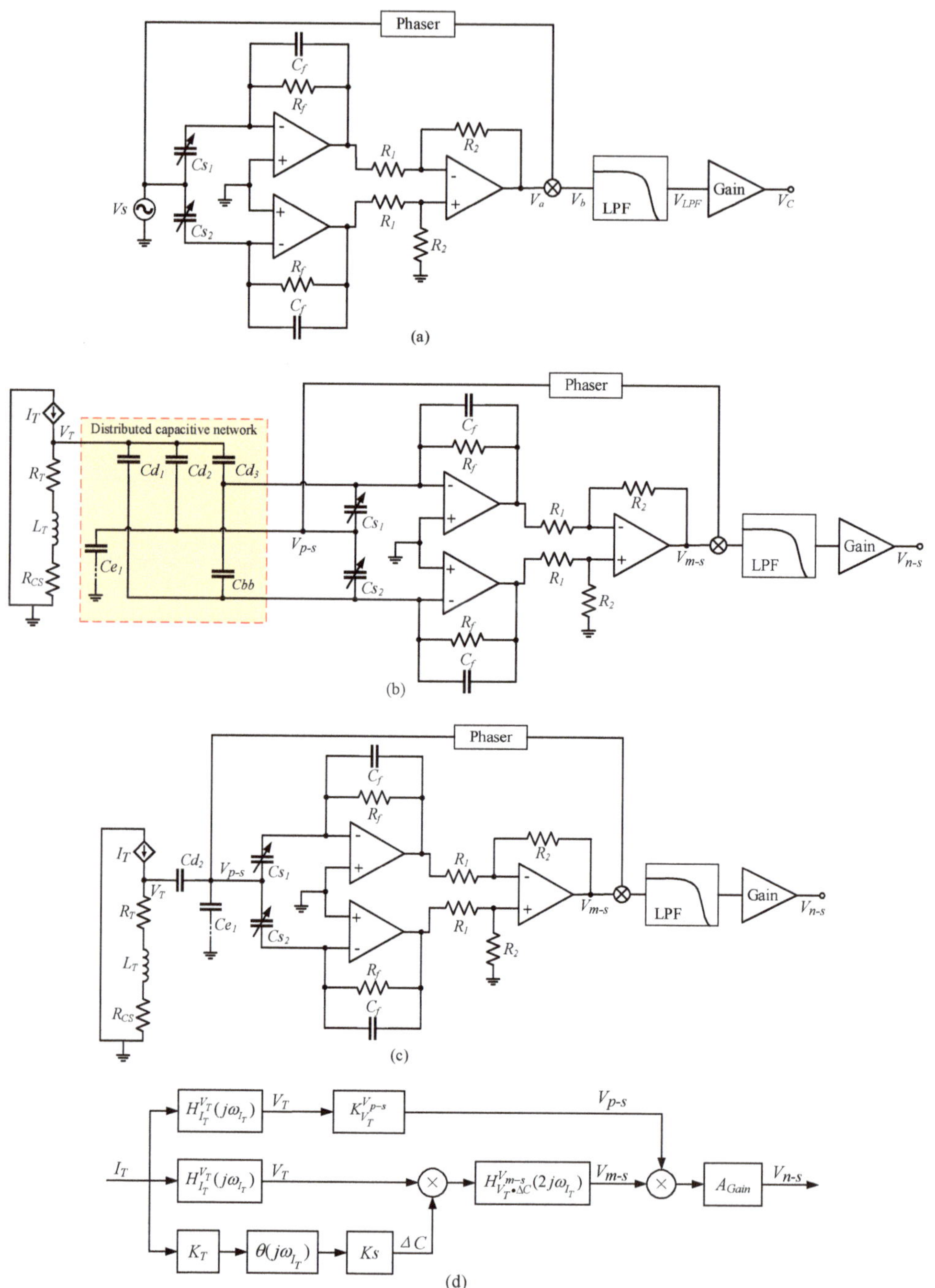

Figure 4. SCM detection circuit and its detection noise transfer path. (**a**) SCM detection circuit; (**b**) Detection noise transfer path composed of SCM detection circuit; (**c**) Simplified path of detection noise transfer formed by SCM detection circuit; (**d**) Structure of detection noise transfer system in SCM detection circuit.

The modulated signal V_a can be expressed as:

$$V_a = \frac{C_{s2} - C_{s1}}{\frac{1}{j\omega R_f} + C_f} \frac{R_2}{R_1} V_s = \frac{2\Delta C A_{diif}}{\frac{1}{j\omega R_f} + C_f} V_s \tag{4}$$

where $A_{diif} = \frac{R_2}{R_1}$ is the amplification factor of the differential amplifier circuit. Generally, the value of R_f ranges from dozens to hundreds of MΩ; thus, Equation (4) can be simplified as:

$$V_a = \frac{2A_{diif}}{C_f} \Delta C V_s \tag{5}$$

It is assumed that sin carrier signal $V_s = A_s \sin(\omega_s t)$ and differential capacitance variation $\Delta C = A_c \sin(\omega_c t)$, where A_s and A_c are amplitude V_s and ΔC, respectively; ω_s is the frequency of carrier signal V_s, usually from tens of kHz to tens of MHz; and ω_c is the effective detectable frequency of ΔC, which is related to the bandwidth of the accelerometer closed-loop system, generally less than 1 kHz. We take the time-domain representation of V_s and ΔC into Equation (5) to obtain:

$$V_a = -\frac{A_{diif}}{C_f} A_c A_s [\cos((\omega_s + \omega_c)t) - \cos((\omega_s - \omega_c)t)] \tag{6}$$

The modulated signal V_a is mixed with the phase-adjusted carrier signal $V_{s+\varphi} = A_s \sin(\omega_s t + \varphi)$ through the mixer, and the mixing output signal V_b can be expressed as:

$$V_b = -\frac{A_{diif}}{2C_f} A_c A_s^2 [\sin((2\omega_s + \omega_c)t + \varphi) - \sin(\omega_c t - \varphi)] \tag{7}$$

There are high-frequency signals of frequency $2\omega_s + \omega_c$ and low-frequency signals of the same frequency ΔC in V_b. When V_b passes through a low-pass filter with a cut-off frequency of ω_c, the high-frequency signals in V_b are filtered out, and only the low-frequency signals in the same frequency as ΔC are retained. The filter output signal V_{LPF} can be expressed as:

$$V_{LPF} = \frac{A_{diif} A_s^2}{2C_f} A_c \sin(\omega_c t - \varphi) \tag{8}$$

After V_{LPF} adjusts the gain through the amplifier, the output voltage V_C of the SCM detection circuit is obtained:

$$V_C = \frac{A_{diif} A_{Gain} A_s^2}{2C_f} \Delta C \tag{9}$$

where A_{Gain} is the amplification factor of the amplifier circuit. According to Equation (9), the output voltage V_C of the SCM detection circuit is directly proportional to the variation of differential capacitance ΔC.

According to the equivalent circuit model of electric field coupling detection noise and the simulation results of distributed capacitance, the detection noise transfer path formed by the SCM detection circuit is as shown in Figure 4b. After the voltage V_T on the torquer coil passes through the distributed capacitance network, it is loaded onto the three plates of the differential capacitor, and then passed into the charge amplification circuit and the differential amplification circuit to form a pseudo-modulated signal V_{m-s}. The signal loaded on the fixed plate is also passed into the phase-shift circuit as a pseudo-carrier signal V_{p-s}. After mixing V_{p-s} with V_{m-s}, the electric field coupling detection noise V_{n-s} is formed through a low-pass filter and gain-adjustment circuit. In the SCM detection noise transfer path, V_T forms a common-mode signal when it is transmitted to the detection circuit through C_{d1} and C_{d3}, which is cancelled when it goes through differential amplification, and does not affect the differential capacitance detection. C_{bb} does not divide voltage or affect differential capacitance detection. We remove C_{d1}, C_{d3}, and C_{bb} to obtain a simplified detection noise transfer path of the SCM detection circuit, as shown in Figure 4c. In the transfer path of the detection noise of the SCM detection circuit, the distributed capacitance C_{d2} between the torquer coil and the yoke iron and C_{e1} between the yoke iron and the shell play a major role.

The process in which the torquer drive current I_T generates voltage V_T from the torquer coil can be expressed as:

$$H_{I_T}^{V_T}(j\omega_{I_T}) = \frac{V_T}{I_T} = R_T + R_{CS} + j\omega_{I_T} \tag{10}$$

where ω_{I_T} is the frequency of the torquer drive current I_T. When the accelerometer float shell is used, the pseudo-carrier signal V_{p-sF} can be expressed as:

$$V_{p-sF} = \frac{C_{d2}}{C_{d2} + C_{s1} + C_{s2}} V_T = \frac{C_{d2}}{C_{d2} + 2C_0} V_T \tag{11}$$

The pseudo-modulated signal V_{m-sF} can be expressed as:

$$V_{m-sF} = \frac{2j\omega_{I_T} R_f}{1 + 2j\omega_{I_T} R_f C_f}\left(\frac{C_{d2}C_{s2}}{C_{d2} + C_{s2}} - \frac{C_{d2}C_{s1}}{C_{d2} + C_{s1}}\right)\frac{R_2}{R_1} V_T = \frac{2j\omega_{I_T} R_f}{1 + 2j\omega_{I_T} R_f C_f}\frac{2C_{d2}^2 \Delta C}{(C_{d2} + C_0)^2 - \Delta C^2}\frac{R_2}{R_1} V_T \tag{12}$$

In Equation (12), $\Delta C^2 \ll (C_{d2} + C_0)^2$ can be ignored, and V_{m-sF} can be expressed as:

$$V_{m-sF} = \frac{2j\omega_{I_T} R_f}{1 + 2j\omega_{I_T} R_f C_f}\frac{2C_{d2}^2}{(C_{d2} + C_0)^2}\frac{R_2}{R_1} V_T \Delta C \tag{13}$$

When the shell is grounded, the pseudo-carrier signal V_{p-sG} can be expressed as:

$$V_{p-sG} = \frac{C_{d2}}{C_{d2} + C_{s1} + C_{s2} + C_{e1}} V_T = \frac{C_{d2}}{C_{d2} + 2C_0 + C_{e1}} V_T \tag{14}$$

The pseudo-modulated signal V_{m-sG} can be expressed as:

$$V_{m-sG} = \frac{2j\omega_{I_T} R_f}{1 + 2j\omega_{I_T} R_f C_f}\left(\frac{C_{d2}C_{s2}}{C_{d2} + C_{s2} + C_{e1}} - \frac{C_{d2}C_{s1}}{C_{d2} + C_{s1} + C_{e1}}\right)\frac{R_2}{R_1} V_T \approx \frac{2j\omega_{I_T} R_f}{1 + 2j\omega_{I_T} R_f C_f}\frac{2C_{d2}^2}{(C_{d2} + C_{e1} + C_0)^2}\frac{R_2}{R_1} V_T \Delta C \tag{15}$$

The accelerometer closed-loop system is linear; I_T, V_T, and ΔC have the same frequency; and the maximum frequency is equal to the closed-loop system bandwidth. V_T and ΔC multiply to produce a double-frequency effect, so that the frequency of V_{m-s} is twice that of I_T, and then through the mixing step, the output-signal frequency is three times that of I_T. Generally, the cut-off frequency of the low-pass filter circuit of the carrier modulation detection circuit is several times that of the closed-loop system bandwidth; therefore, the attenuation of the frequency-doubling signal through the filter circuit can be ignored. The structure of the detection noise transfer system of the SCM detection circuit is as shown in Figure 4d. The detection noise V_{n-s} of SCM detection circuit can be expressed as:

$$V_{n-s} = A_{Gain} K_{V_T}^{V_{p-s}} H_{V_T \cdot \Delta C}^{V_{m-s}}(2j\omega_{I_T})\left(H_{I_T}^{V_T}(j\omega_{I_T})\right)^2 K_T \theta(j\omega_{I_T}) K_s I_T^3 \tag{16}$$

where $K_{V_T}^{V_{p-s}}$ represents the transfer process from V_T to carrier signal V_{p-s}, $K_{V_T}^{V_{p-s}} = \frac{C_{d2}}{C_{d2}+2C_0}$ when the float shell is used, and $K_{V_T}^{V_{p-s}} = \frac{C_{d2}}{C_{d2}+2C_0+C_{e1}}$ when the shell is grounded; $H_{V_T \cdot \Delta C}^{V_{m-s}}(j\omega_{I_T})$ represents the transfer process from the product of V_T and ΔC to V_{m-s}. $H_{V_T \cdot \Delta C}^{V_{a-s}}(j\omega_{I_T}) = \frac{2j\omega_{I_T} R_f}{1+2j\omega_{I_T} R_f C_f}\frac{2C_{d2}^2}{(C_{d2}+C_0)^2}\frac{R_2}{R_1}$ when the float shell is used, and $H_{V_T \cdot \Delta C}^{V_{m-s}}(2j\omega_{I_T}) = \frac{2j\omega_{I_T} R_f}{1+2j\omega_{I_T} R_f C_f}\frac{2C_{d2}^2}{(C_{d2}+C_{e1}+C_0)^2}\frac{R_2}{R_1}$ when the shell is grounded.

3.2. Detection Noise Transfer System Structure of DCM Detection Circuit

When a DCM detection circuit is used to detect differential capacitance, the structure of the detection circuit is as shown in Figure 5a. V_s and $-V_s$ are high-frequency sinusoidal carrier signals with the same frequency and amplitude and opposite phase, which are separately input to the two movable plates of the differential capacitor. After the output signals of the movable plate pass through the charge amplifier circuit, the modulated signal V_a can be expressed as:

$$V_a = \frac{C_{s2} - C_{s1}}{\frac{1}{j\omega R_f} + C_f} V_s = \frac{2\Delta C}{\frac{1}{j\omega R_f} + C_f} V_s \approx \frac{2}{C_f} \Delta C V_s \quad (17)$$

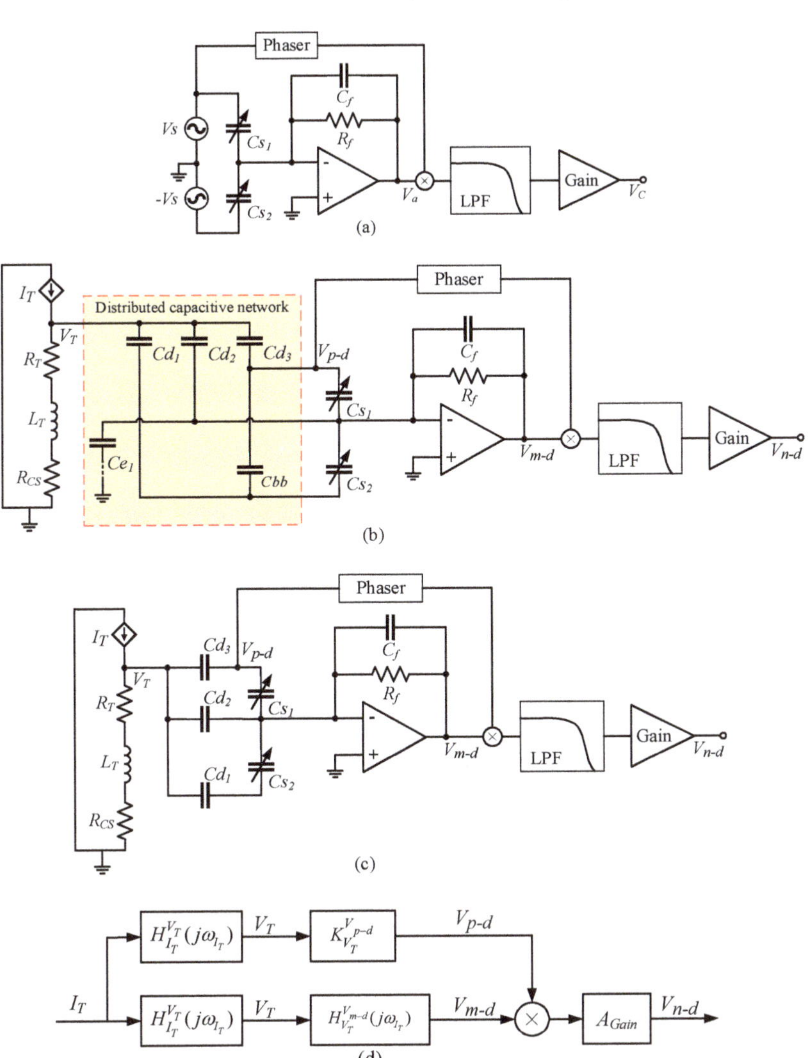

Figure 5. DCM detection circuit and its detection noise transfer path. (**a**) DCM detection circuit; (**b**) Detection noise transfer path composed of DCM detection circuit; (**c**) Simplified path of detection noise transfer formed by DCM detection circuit; (**d**) Structure of detection noise transfer system in DCM detection circuit.

As with the SCM detection circuit, after mixing, filtering, and gain adjustment of the modulated signal V_a are performed, a voltage signal V_C proportional to the change in differential capacitance ΔC can be obtained, and expressed as:

$$V_C = \frac{A_{Gain} A_s^2}{2C_f} \Delta C \tag{18}$$

The detection noise transfer path formed by the DCM detection circuit is shown in Figure 5b. After the voltage V_T on the torquer coil passes through the distributed capacitance network, it is loaded onto the three plates of the differential capacitor, and then passed into the charge amplification circuit to form a pseudo-modulated signal V_{m-d}. The signal loaded on the top plate is also passed into the phase-shift circuit as a pseudo-carrier signal V_{p-d}. After mixing V_{p-d} with V_{m-d}, the electric field coupling detection noise V_{n-d} is formed through a low-pass filter and gain adjustment circuit. In the DCM detection noise transfer path, C_{e1} is connected to the reverse port of the operational amplifier, which is approximately grounded and does not affect the differential capacitance detection. C_{bb} does not divide the voltage or affect differential capacitance detection. We remove C_{e1} and C_{bb} to obtain a simplified detection noise transfer path of the DCM detection circuit, as shown in Figure 5c. In the transfer path of the detection noise of the DCM detection circuit, the distributed capacitances C_{d1}, C_{d2}, and C_{d3} between the torquer coil and the three plates of the differential capacitor play a major role.

The pseudo-carrier signal V_{p-d} can be expressed as

$$V_{p-d} = \frac{C_{d3}}{C_{d3} + C_{s1}} V_T = \frac{C_{d3}}{C_{d3} + C_0 - \Delta C} V_T \approx \frac{C_{d3}}{C_{d3} + C_0} V_T \tag{19}$$

The equivalent capacitance C_{in} of the capacitance network composed of distributed capacitances C_{d1}, C_{d2}, and C_{d3} and differential capacitors C_{s1} and C_{s2} can be expressed as

$$C_{in} = \frac{C_{d3} C_{s1}}{C_{d3} + C_{s1}} + \frac{C_{d1} C_{s2}}{C_{d3} + C_{s2}} + C_{d2} = \frac{2C_{d3}(C_0^2 - \Delta C^2) + 2C_{d3}^2 C_0}{(C_{d3} + C_0)^2 - \Delta C^2} + C_{d2} \approx 2\frac{C_{d3} C_0}{C_{d3} + C_0} + C_{d2} \tag{20}$$

The pseudo-modulated signal V_{m-d} can be expressed as

$$V_{m-d} = -\frac{j\omega_{I_T} R_f C_{in}}{1 + j\omega_{I_T} R_f C_f} V_T \tag{21}$$

The frequency of V_{p-d} and V_{m-d} is the same, and the frequency of the mixing output signal V_{n-d} is twice that of I_T. The structure of the detection noise transfer system of the DCM detection circuit is shown in Figure 5d. The detection noise V_{n-d} of the DCM detection circuit can be expressed as:

$$V_{n-d} = A_{Gain} K_{V_T}^{V_{p-d}} H_{V_T}^{V_{m-d}}(j\omega_{I_T}) \left(H_{I_T}^{V_T}(j\omega_{I_T}) \right)^2 I_T^2 \tag{22}$$

where $K_{V_T}^{V_{p-d}} = \frac{C_{d3}}{C_{d3}+C_0}$ represents the transfer process from V_T to carrier signal V_{p-d}, and $H_{V_T}^{V_{m-d}}(j\omega_{I_T}) = -\frac{j\omega_{I_T} R_f C_{in}}{1+j\omega_{I_T} R_f C_f}$ represents the transfer process from V_T to V_{m-d}.

4. Analysis and Experiment in Relation to Detection Noise Characteristics

4.1. Analysis of Detection Noise Transfer in The Closed-Loop System of The QFA

The detection noise V_n is superimposed onto the output signal V_C of the differential capacitance detection circuit and enters the closed-loop system of the QFA. The introduction position of the detection noise V_n is shown in Figure 6. The transfer function between V_n and the drive current sampling output V_0 is

$$H_{V_n}^{V_0}(s) = \frac{M(s)K_I K_U}{1 - M(s)K_I K_T \theta(s) K_s K_a} \tag{23}$$

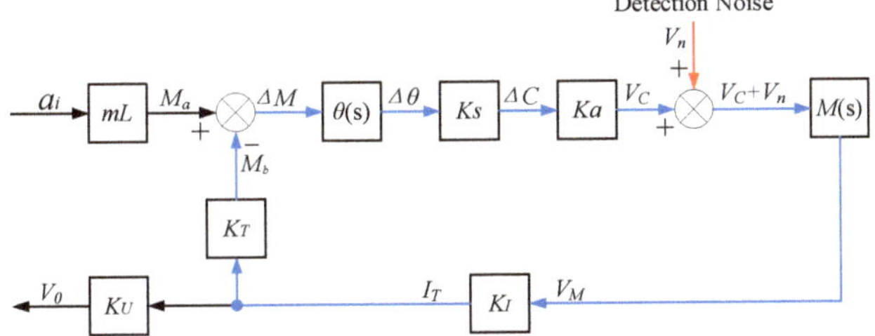

Figure 6. The introduction position of detection noise in the closed-loop system.

According to Equations (12), (21), and (22), when the SCM detection circuit is used to detect differential capacitance, the transfer relationship between the equivalent acceleration a_{n-s} of the detected noise and input acceleration a_i can be expressed as:

$$a_{n-s} = K_{V_0}^{a_n} H_{V_n}^{V_0}(\omega_{a_i}) V_{n-s} = K_{V_0}^{a_n} H_{V_n}^{V_0}(j\omega_{a_i}) H_{I_T^3}^{V_{n-s}}(j\omega_{a_i}) (H(j\omega_{a_i}))^3 a_i^3 \tag{24}$$

where ω_{a_i} is the frequency of input acceleration a_i, and $K_{V_0}^{a_n}$ is the transfer function between current sampling output V_0 and the equivalent acceleration a_n of detection noise, equal to the reciprocal product of sampling circuit transfer function K_U and accelerometer scale factor K_1. $H_{I_T^3}^{V_{n-s}}(j\omega_{a_i})$ represents the transfer relationship between torquer converter drive current I_T and detection noise V_{n-s}.

When a DCM detection circuit is used to detect differential capacitance, the transfer relationship between equivalent acceleration a_{n-d} of the detected noise and input acceleration a_i can be expressed as:

$$a_{n-d} = K_{V_0}^{a_n} H_{V_n}^{V_0}(\omega_a) V_{n-d} = K_{V_0}^{a_n} H_{V_n}^{V_0}(j\omega_a) H_{I_T^2}^{V_{n-d}}(j\omega_a) (H(j\omega_a))^2 a_i^2 \tag{25}$$

where $H_{I_T^2}^{V_{n-d}}(j\omega_{a_i})$ represents the transfer relationship between the torquer drive current I_T and detection noise V_{n-d}.

4.2. Analysis of Influencing Factors of Detection Noise

4.2.1. Analysis of Influencing Factors of Detection Noise in SCM Detection Circuit

According to Equation (24), the magnitude of the detection noise equivalent acceleration a_{n-s} of the SCM detection circuit is determined by $H(s)$, $H_{V_n}^{V_0}(s)$, $H_{I_T^3}^{V_{n-s}}(s)$, and $K_{V_0}^{a_n}$. Under the condition that the structure of the accelerometer pendulum reed and the bandwidth of the closed-loop system are constant, K_B, $\theta(s)$, K_s, K_T, $H(s)$, and $K_{V_0}^{a_n}$ are unchanged, and the product of $\theta(s)K_s K_a M(s) K_I K_T$ is fixed. If K_I and K_a are adjusted, $M(s)$ will be adjusted in equal proportion. Adjusting K_I only affects the transfer function $H_{V_n}^{V_0}(s)$, but K_I in the numerator and denominator of $H_{V_n}^{V_0}(s)$ is multiplied by $M(s)$; thus, adjusting K_I does not change the value of the transfer function $H_{V_n}^{V_0}(s)$, and cannot suppress a_{n-s}.

According to Equation (9), the transfer function K_a of the SCM detection circuit can be expressed as:

$$K_a = \frac{A_{diif} A_{Gain} A_s^2}{2C_f} \tag{26}$$

K_a is proportional to A_{diif}, A_{Gain}, and A_S squared, and inversely proportional to C_f. The changes in A_{diif}, A_{Gain}, and C_f affect not only K_a but also the transfer function $H_{I_T^3}^{V_{n-s}}(s)$. According to Equations (23), (24), and (26), when the accelerometer float shell is used, the equivalent acceleration of detection noise a_{n-sF} is

$$a_{n-sF} = K_{V_0}^{a_n} \frac{K_a M(s) K_I K_U K_T \theta(s) K_s}{1 - M(s) K_I K_T \theta(s) K_s K_a} \frac{2sR_f}{\frac{1}{C_f} + 2sR_f} \frac{2}{A_s^2} \frac{2C_{d2}^2}{(C_{d2} + C_0)^2} K_{V_T}^{V_{p-s}} \left(H_{I_T}^{V_T}(s)\right)^2 (H(s))^3 a_i^3 \tag{27}$$

when the shell is grounded, the equivalent acceleration of detection noise is

$$a_{n-sG} = K_{V_0}^{a_n} \frac{K_a M(s) K_I K_U K_T \theta(s) K_s}{1 - M(s) K_I K_T \theta(s) K_s K_a} \frac{2sR_f}{\frac{1}{C_f} + 2sR_f} \frac{2}{A_s^2} \frac{2C_{d2}^2}{(C_{d2} + C_{e1} + C_0)^2} K_{V_T}^{V_{p-s}} \left(H_{I_T}^{V_T}(s)\right)^2 (H(s))^3 a_i^3 \tag{28}$$

According to Equations (27) and (28), increasing the amplitude of carrier signal V_s and reducing the value of feedback capacitance C_f can inhibit the detection of a_{n-s}.

4.2.2. Analysis of Influencing Factors of Detection Noise in DCM Detection Circuit

According to Equation (24), the magnitude of the detection noise equivalent acceleration a_{n-d} of the DCM detection circuit is determined by $H(s)$, $H_{V_n}^{V_0}(s)$, $H_{I_T^2}^{V_{n-d}}(s)$, and $K_{V_0}^{a_n}$. As in the case of the single-carrier modulation detection circuit, adjusting K_I cannot suppress a_{n-d}.

According to Equation (18), the transfer function K_a of the DCM detection circuit can be expressed as:

$$K_a = \frac{A_{Gain} A_s^2}{2C_f} \tag{29}$$

According to Equations (23), (25), and (29), the equivalent acceleration of detection noise a_{n-d} is

$$a_{n-d} = -K_{V_0}^{a_n} \frac{K_a M(s) K_I K_U}{1 - M(s) K_I K_T \theta(s) K_s K_a} K_{V_T}^{V_{p-d}} \frac{sR_f C_{in}}{\frac{1}{C_f} + sR_f} \frac{2}{A_s^2} \left(H_{I_T}^{V_T}(s)\right)^2 (H(s))^2 a_i^2 \tag{30}$$

Increasing the amplitude of carrier signal V_s and reducing the value of feedback capacitance C_f can inhibit the detection of a_{n-d}.

Through the analysis of the factors influencing the detection noise of the above two detection circuits, it can be seen that when the structure of the accelerometer pendulum reed and the bandwidth of the closed-loop system are constant, the factors influencing the detection noise of the two detection circuits are consistent, and the equivalent acceleration of the detection noise via electric field coupling can be suppressed by increasing the amplitude of the carrier signal of the detection circuit V_s and reducing the value of the feedback capacitor C_f.

4.3. Experiment in Respect of Detection Noise Characteristics

In this study, the transfer characteristic and influencing factors of electric field coupling detection noise were analyzed through simulation. The torquer coil parameters of the typical QFA studied are $L_T = 31.2$ mH, $R_T = 424\ \Omega$, $R_{CS} = 1\ \Omega$. The typical parameters of the SCM and DCM detection circuits are $R_1 = 1\ \mathrm{K\Omega}$, $R_2 = 1\ \mathrm{K\Omega}$, $R_f = 100\ \mathrm{M\Omega}$, $C_f = 20$ pF and $A_{Gain} = 2.4$. The frequency of sinusoidal carrier signal V_s is 50 kHz, the amplitude is 5 V and the typical value of K_a is 1.5 V/pF. Typical parameters of the QFA closed-loop system in Figure 2 are as follows: $K_B = 6.3 \times 10^{-6}$ kg · m, $J = 1.1 \times 10^{-8}$ kg · m^2, $C = 2.0 \times 10^{-4}$ N · m · s/rad, $K = 2.2 \times 10^{-3}$ N · m/rad, $K_s = 12{,}000$ pF/rad, $K_T = 6.3 \times 10^{-6}$ N · m/A, $K_U = 1$ V/A, and $K_I = 10$ mA/V. $M(s)$ represents proportional feedback to meet the closed-loop bandwidth requirements, and the proportional feedback gain $K_M = 0.3$. The accelerometer scale factor $K_1 = 1$ mA/g, the acceleration measurement range is ±10 g, the system bandwidth is 300 Hz, and the effective resolution is 50 μg.

4.3.1. Experiment on the Transfer Characteristics of Noise-Detection

The amplitude-frequency characteristics of $H(s)$ and $H_{V_n}^{V_0}(s)$ are shown in Figure 7. The cut-off frequency of the closed-loop transfer function of the accelerometer is 300 Hz, which is consistent with the actual bandwidth of the system. $H_{V_n}^{V_0}(s)$ monotonically increases within the range of three times the input acceleration, and the transmission of detection noise in the closed-loop system is a high-pass characteristic.

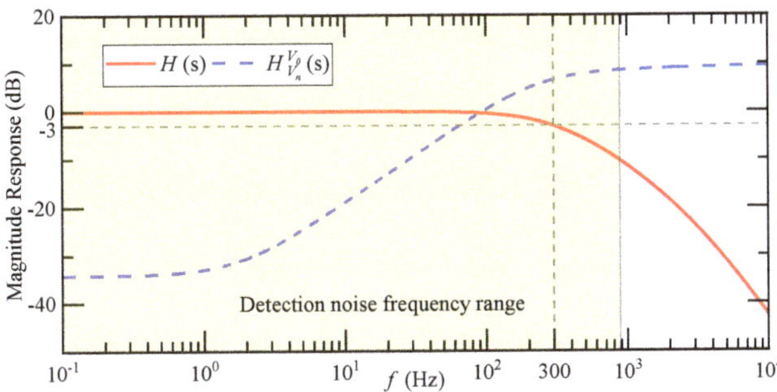

Figure 7. Amplitude-frequency characteristics of $H(s)$ and $H_{V_n}^{V_0}(s)$.

When the system input acceleration is 1 g, the relationship between the detected noise equivalent acceleration a_n and the input acceleration a_i frequency is as shown in Figure 8. a_{n-sF} is the detection noise equivalent acceleration of the SCM detection circuit when the float shell is used, a_{n-sG} is the detection noise equivalent acceleration of the SCM detection circuit when the shell is grounded and a_{n-d} is the detection noise equivalent acceleration of the DCM detection circuit. The magnitude of a_n is positively correlated with the frequency of a_i. The higher the frequency of a_i, the greater the equivalent acceleration of detection noise. The transfer process from a_i to a_n shows a high-pass characteristic. The equivalent acceleration of detection noise of the SCM detection circuit reaches the maximum when the a_i frequency is 70 Hz, and the equivalent acceleration of detection noise of the DCM detection circuit reaches the maximum when the a_i frequency is 220 Hz.

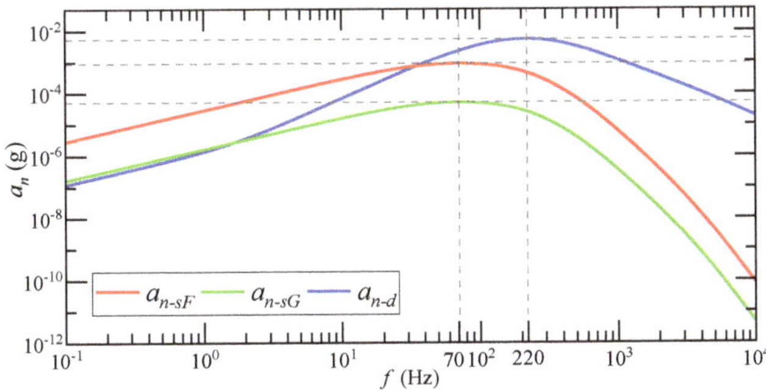

Figure 8. The relationship between the magnitude of a_n and the frequency of a_i under 1 g input acceleration.

The system inputs a random acceleration of 0~10 g and a frequency of 0~300 Hz, and the equivalent acceleration of detection noise of each type of carrier modulation detection circuit is shown in Figure 9. The mean of a_{n-sF} is 10.3 μg, a_{n-sG} is 0.2 μg, and a_{n-d} is 41.7 μg.

The equivalent acceleration of the detection noise is greater when the DCM detection circuit is used than when the SCM detection circuit is used. When an SCM detection circuit is used, the shell grounding can effectively reduce the equivalent acceleration of the detection noise.

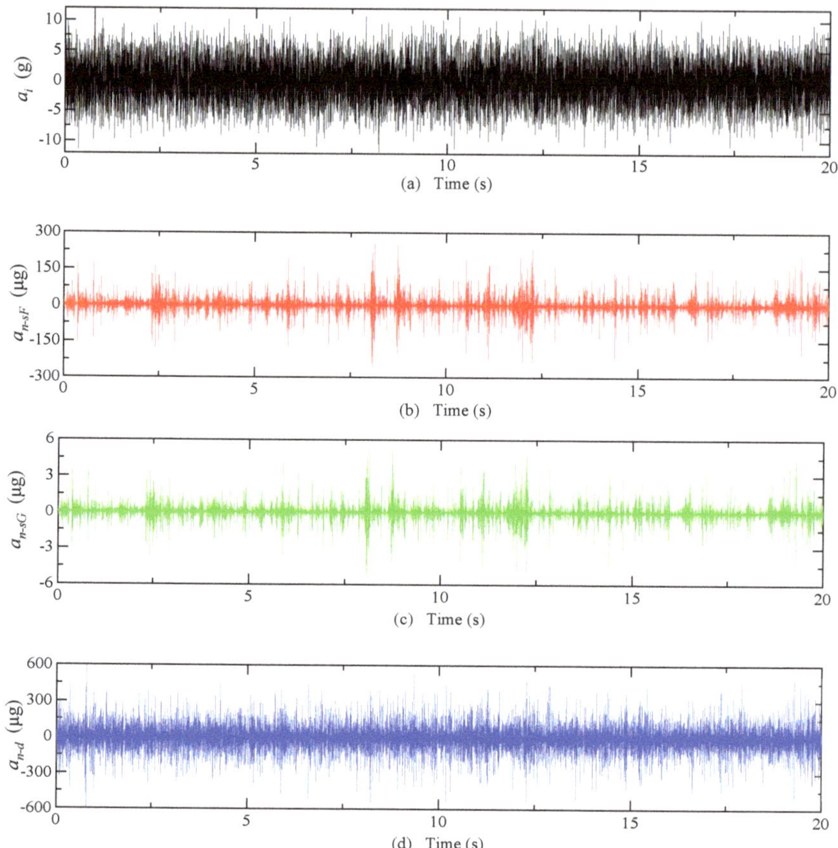

Figure 9. Electric field coupling detection noise simulation results. (**a**) Input acceleration; (**b**) The detection noise equivalent acceleration of the SCM detection circuit when the float shell is used; (**c**) The detection noise equivalent acceleration of the SCM detection circuit when the shell is grounded; (**d**) The detection noise equivalent acceleration of the DCM detection circuit.

The comparison of different types of noise values in the QFA system is shown in Table 7. The above analysis shows that the detection noise equivalent acceleration is close to the effective resolution of the accelerometer of 50 μg, which is a factor that cannot be ignored, as doing so affects the measurement accuracy of the QFA.

Table 7. Comparison of different types of noise values in QFA system.

Noise Type	Value (μg)
Mechanical thermal noise	1.3×10^{-5}
Bias repeatability	30
Detect circuit noise	1.53
Electric field coupling detection noise	41.7

4.3.2. Experiment on Influencing Factors of Detection Noise

In this section, the simulation analysis results for the factors affecting the detection noise are given in the case where the accelerometer pendulum structure and the closed-loop system bandwidth are constant. Within the allowable adjustment range of conventional carrier modulation detection circuit parameters, when V_s and C_f are adjusted, the change trend of K_a and K_M is as shown in Figure 10. K_a increases with the increase in V_s and decreases with the increase in C_f. The change trend of K_M is the opposite. The product of K_a and K_M remains unchanged at 0.45, the transfer function of the accelerometer closed-loop system remains unchanged, and the system bandwidth remains unchanged at 300 Hz. Therefore, increasing the carrier signal V_s magnitude and decreasing the feedback capacitance C_f will not affect the closed-loop system bandwidth.

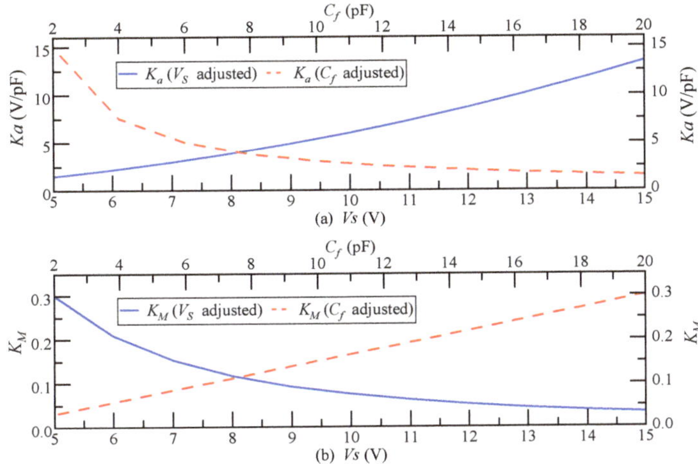

Figure 10. (**a**) The changing trend of K_a when V_s and C_f are adjusted; (**b**) the changing trend of K_M when V_s and C_f are adjusted.

The influence of increasing V_s on the amplitude-frequency characteristic curve of a_n is shown in Figure 11. V_s doubled, and the amplitude-frequency characteristic curve of a_n decreases by 12 dB. At this time, the mean of a_{n-sF} is 2.58 μg, a_{n-sG} is 0.05 μg, and a_{n-d} is 10.4 μg, while a_n attenuates 12 dB. Increasing the magnitude of carrier signal V_s can suppress the detection noise.

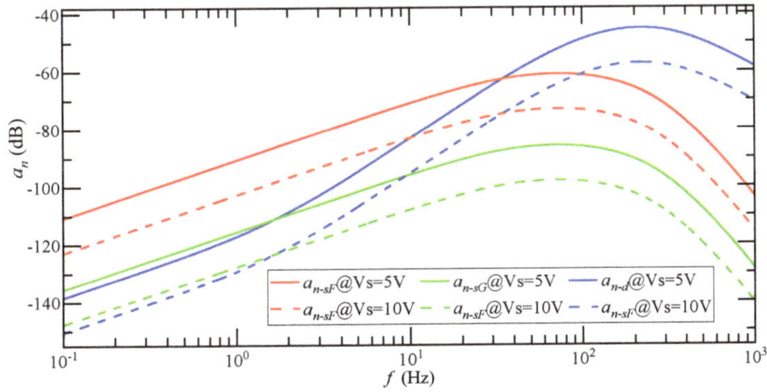

Figure 11. Amplitude-frequency characteristic curve of a_n when V_s is doubled.

The influence of reducing C_f on the amplitude-frequency characteristic curve of a_n is shown in Figure 12. When the input acceleration frequency is less than 20 Hz, C_f reduces by half, and the amplitude-frequency characteristic curve decreases by 6 dB. When the input acceleration frequency is greater than 20 Hz, the attenuation of the amplitude-frequency characteristic curve gradually decreases, and when it reaches more than 300 Hz, it is consistent with the original amplitude-frequency characteristic curve. When C_f is reduced by one time, the mean of a_{n-sF} is 8.48 μg, a_{n-sG} is 0.2 ug, and a_{n-d} is 35.3 μg, while the average attenuation of a_n is 1.6 dB. Reducing the feedback capacitance C_f of the detection circuit can suppress the detection noise. The effect of reducing C_f on the low-frequency component of a_n was better, but the effect on the high-frequency component was not obvious.

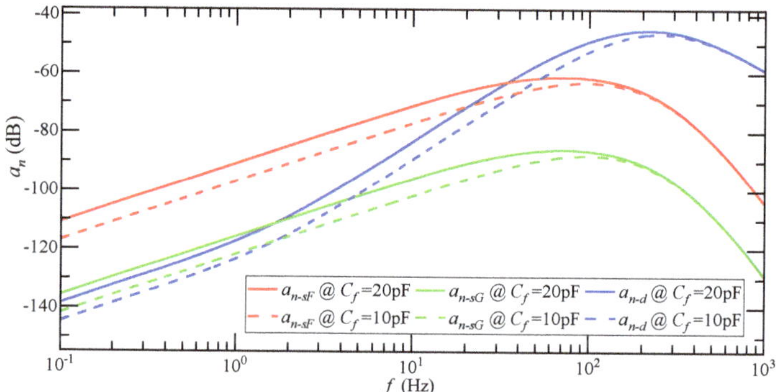

Figure 12. Amplitude-frequency characteristic curve of a_n when C_f decreases by one time.

When V_s and C_f are adjusted, the change rule of the mean magnitude of a_n is as shown in Figure 13. Increasing V_s is more effective than decreasing C_f; when $V_s = 6.55$ V and $C_f = 4.8$ pF, the inhibition effect on a_n is the same.

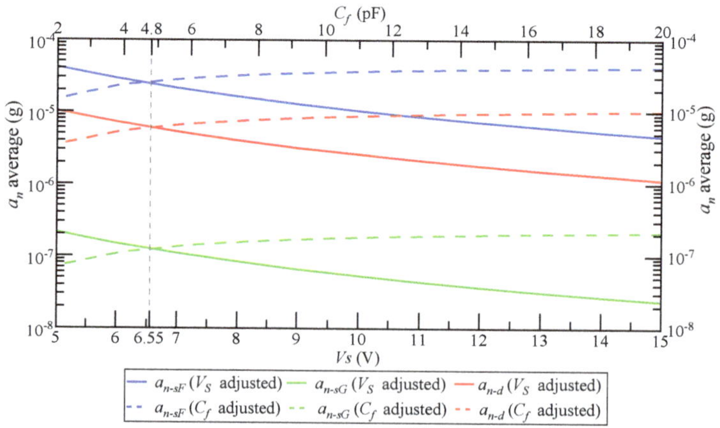

Figure 13. The change rule of the average magnitude of a_n when V_s and C_f are adjusted.

When V_s is adjusted to the maximum magnitude and C_f to the minimum value, the average magnitude of a_n is as shown in Table 8. At the same time, when V_s is a maximum and C_f is a minimum, the suppression effect of the detection noise can be maximized.

Table 8. Average value of a_n for different values of V_s and C_f.

V_S(V)	C_f (pF)	a_{n-sF}Average (μg)	a_{n-sG}Average (μg)	a_{n-d}Average (μg)
15.0	20.0	1.1	0.024	4.6
5.0	2.0	3.4	0.071	14.8
15.0	2.0	0.4	0.0079	1.6

The relationship between the average a_n and V_s can be approximately expressed as

$$a_{n-avg} = k_V V_s^{-2} \tag{31}$$

where a_{n-avg} is the average value of a_n, and k_V is the gain coefficient. When the SCM detection circuit is adopted for the float shell, $k_V = 2.6 \times 10^{-4}$; when the SCM detection circuit is adopted for the grounded shell, $k_V = 5.4 \times 10^{-6}$; and when the DCM detection circuit is adopted, $k_V = 1.1 \times 10^{-3}$.

The relationship between the average a_n and C_f can be approximately expressed as

$$a_{n-avg} = k_f \log\left(C_f\right) + k_0 \tag{32}$$

where k_f is the gain coefficient and k_0 is the bias coefficient. When the SCM detection circuit is adopted for the float shell, $k_f = 7.0 \times 10^{-6}$, $k_0 = 1.3 \times 10^{-6}$; when the SCM detection circuit is adopted for the grounded shell, $k_f = 1.5 \times 10^{-7}$, $k_0 = 2.7 \times 10^{-5}$; and when the DCM detection circuit is adopted, $k_f = 2.7 \times 10^{-5}$, $k_0 = 7.5 \times 10^{-6}$.

When designing a QFA with a_n accuracy higher than 1 μg, it is necessary to keep the average value of a_n below 0.1 μg to ensure that it has a small impact on the QFA accuracy. However, according to the average value of a_n in Table 5, it can be seen that within the allowable adjustment range of the conventional carrier modulation detection circuit parameters, the low noise requirement can be met only when the shell is grounded and the SCM detection circuit is adopted. In the other two cases, no matter how the circuit parameters are adjusted, a_{n-avg} is always greater than 0.1 μg. According to Equations (31) and (32), in order to make a_{n-avg} less than 0.1 μg, V_s should be adjusted to above 51.0 V or C_f should be adjusted to below 0.67 pF when the float shell and the detection circuit are adopted by SCM. When the DCM detection circuit is used, V_s should be adjusted to above 104.9 V or C_f should be adjusted to below 0.53 pF. According to Equations (9), (18), (13), and (21), increasing V_s and C_f improves the signal-to-noise ratio of the detection circuit by increasing the gain of the differential capacitance detection circuit. However, this method does not really reduce the detection noise of electric field coupling, which requires the gain of the differential capacitance detection circuit to be large enough, which is difficult to realize. Therefore, it is necessary to further reduce the detection noise of electric field coupling for a high-accuracy accelerometer.

5. Suppression Method of Electric Field Coupling Detection Noise

5.1. Optimization Analysis of Carrier Modulation Differential Capacitance Detection Circuit

According to the structure of the detection noise transfer system of the two carrier modulation detection circuits, the detection noise V_n is equal to the product of V_m and V_p. In order to truly reduce the detection noise, V_n, V_m, or V_p should be reduced. Increasing the feedback capacitance C_f can reduce V_m, but increasing C_f will increase a_{n-avg}; therefore, V_n can only be reduced by reducing V_p. Reducing V_p can not only reduce the gain from the driving current to the detection-noise-transmission process but will also not cause an $H(s)$, $H_{V_n}^{V_0}(s)$, and $K_{V_0}^{a_n}$ gain change. According to Figures 4b and 5b, V_p only passes through a phase-shifting circuit before mixing with V_m through the multiplier. Since the carrier signal V_S through the phase-shifting circuit is a high-frequency signal, and pseudo-carrier signal V_p is a low-frequency signal within 300 Hz, the value of V_p can be suppressed by adding a high-pass filter in front of the phase-shifting circuit. The optimized carrier-modulated differential capacitance detection circuit is shown in Figure 14.

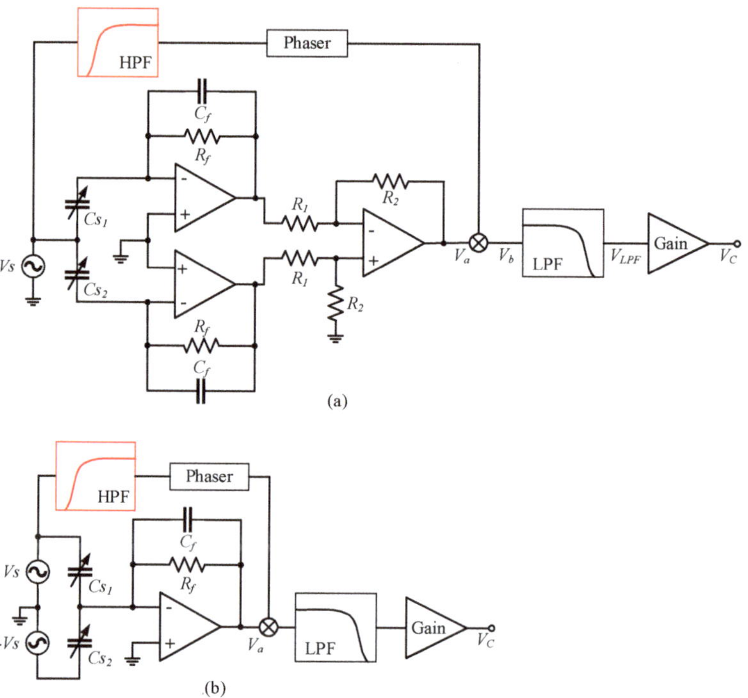

Figure 14. (**a**) The optimized SCM detection circuit; (**b**) The optimized DCM detection circuit.

The design of the second-order Butterworth high-pass filter is shown in Figure 15a. The cut-off frequency of the high-pass filter is 10 kHz, the gain is 0 dB, the stopband frequency is 300 Hz, and the stopband attenuation is −60 dB. The transfer function $H_{HPF}(s)$ can be expressed as

$$\begin{aligned} H_{HPF}(s) &= \frac{R_1R_2C_1C_2s^2}{1+(C_1+C_2)R_2s+R_1R_2C_1C_2s^2} \\ &= \frac{2.54\times10^{-10}s^2}{1+2.26\times10^{-5}s+2.54\times10^{-10}s^2} \end{aligned} \tag{33}$$

The amplitude-frequency characteristic curve of the high-pass filter in Figure 15b shows that the high-pass filter effectively suppresses the pseudo-carrier signal V_p, and the carrier signal V_s is not affected.

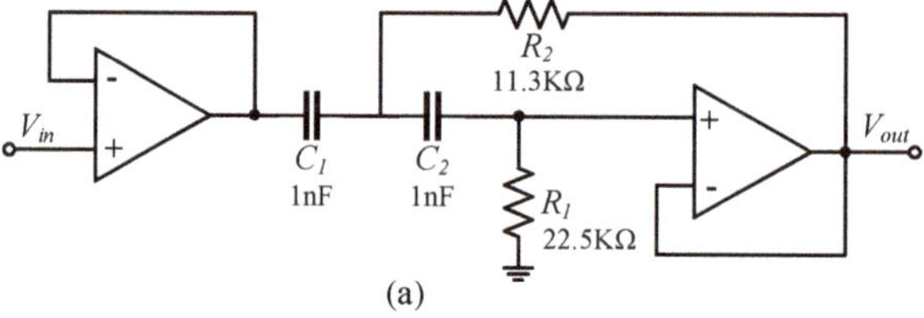

Figure 15. *Cont.*

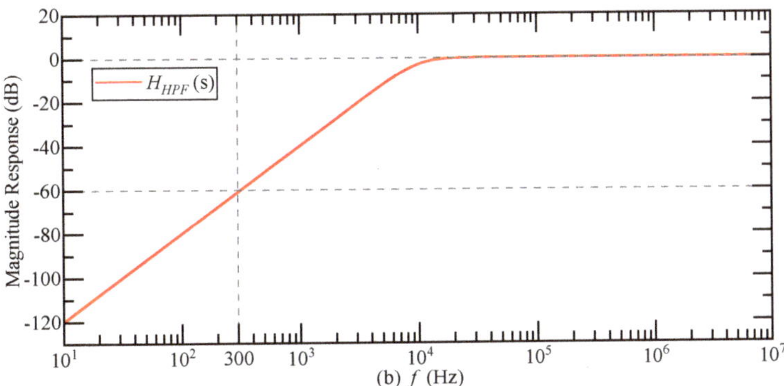

Figure 15. (**a**) High-pass filter structure; (**b**) Amplitude-frequency characteristic curve of high-pass filter.

The structure of the optimized electric field coupling detection noise transfer system is shown in Figure 16.

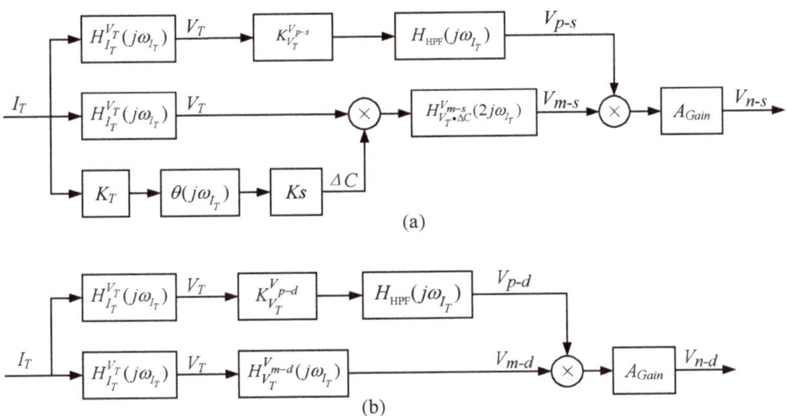

Figure 16. (**a**) Optimized SCM detection circuit noise transfer system structure; (**b**) Optimized DCM detection circuit noise transfer system structure.

5.2. Analysis of Suppression Effect of Electric Field Coupling Detection Noise

When the system input acceleration is 1 g, the relationship between the detected noise equivalent acceleration a_n and the input acceleration a_i frequency before and after the optimization of the differential capacitance detection circuit is as shown in Figure 17. Compared with before optimization, a_n decreases significantly after the addition of a high-pass filter, and attenuates 130 dB on average in the bandwidth of 300 Hz.

The system inputs a random acceleration of 0~10 g and a frequency of 0~300 Hz; the equivalent acceleration of detected noise after optimization is shown in Figure 18. The mean of a_{n-sF} is 1.19×10^{-3} μg, a_{n-sG} is 2.47×10^{-5} μg, and a_{n-d} is 8.47×10^{-3} μg.

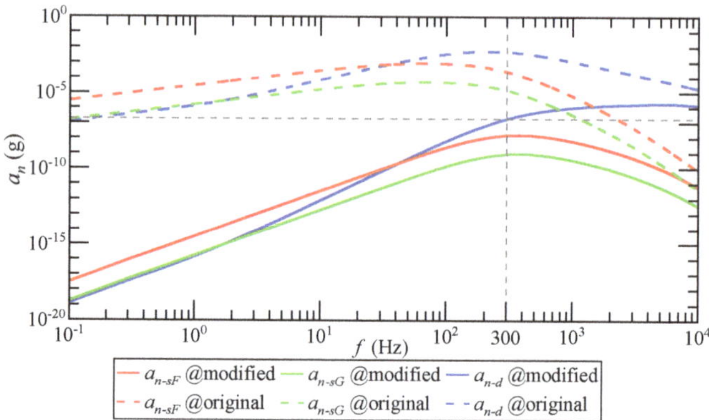

Figure 17. Before and after detection-circuit optimization, the relationship between the magnitude of a_n and the frequency of a_i under 1 g input acceleration.

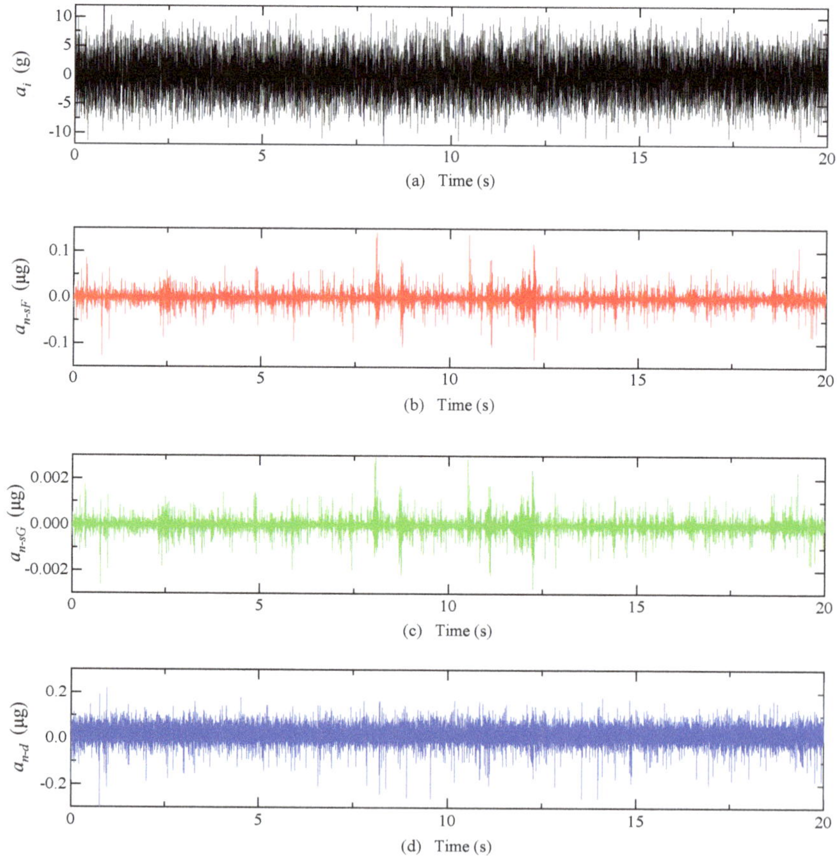

Figure 18. Simulation results of electric field coupling detection noise after optimization. (**a**) Input acceleration; (**b**) The detection noise equivalent acceleration of the SCM detection circuit when the float shell is used; (**c**) The detection noise equivalent acceleration of the SCM detection circuit when the shell is grounded; (**d**) The detection noise equivalent acceleration of the DCM detection circuit.

The comparison of the average value of the equivalent acceleration of the detection noise before and after the optimization of the differential capacitance detection circuit is shown in Table 9. After the detection circuit optimization, the average equivalent acceleration of the detection noise attenuates about 78 dB compared with that before the optimization, which can significantly suppress the electric field coupling detection noise and ensure that the a_{n-avg} is less than 0.1 μg, effectively reducing the impact of the detection noise on the accuracy of the accelerometer.

Table 9. The average value of the detection noise equivalent acceleration before and after the optimization of the differential capacitor detection circuit.

Equivalent Acceleration of Detection Noise	Value Before Optimization (μg)	Optimized Value (μg)	Attenuation (dB)
a_{n-sF} average	10.3	1.19×10^{-3}	78.7
a_{n-sG} average	0.2	2.47×10^{-5}	78.2
a_{n-d} average	41.7	8.47×10^{-3}	73.8

6. Conclusions

This paper focuses on the analysis of the coupling mechanism of the internal electric field of the QFA and the influence of the electric field coupling detection noise on the accuracy of the QFA, establishing the equivalent circuit model of the internal electric field coupling detection noise of the accelerometer. The value of the distributed capacitance inside the accelerometer is simulated to obtain the detection noise transfer system structure of different carrier modulated differential capacitance detection circuits. Through the simulation experiment, the noise of electric field coupling detection is calculated. The experimental results show that the transmission of electric field coupling detection noise in the closed-loop system of the QFA is high-pass characteristic. Within the effective range and bandwidth of the system, the average value of the equivalent acceleration of noise detected by the SCM detection circuit when using the float shell is 10.3 μg; the average value of the equivalent acceleration of noise detected by the SCM detection circuit when the shell is grounded is 0.2 μg; and the average value of the equivalent acceleration of noise detected by the DCM detection circuit is 41.7 μg. The equivalent acceleration of the detection noise is close to the effective resolution of the accelerometer of 50 μg, indicating that the field coupling detection noise is a non-negligible factor affecting the measurement accuracy of the accelerometer. The analysis and experiment on the influence factors of electric field coupling detection noise show that when the structure of the accelerometer pendulum reed and the bandwidth of the closed-loop system are constant, increasing the magnitude of the carrier signal and reducing the feedback capacitance of the detection circuit can reduce the detection noise; however, the suppression degree is limited, and it is difficult to meet the noise requirements of the QFA with 1 μg accuracy. It is necessary to optimize the differential capacitance detection circuit to further reduce the noise of electric field coupling detection. Therefore, a high-pass filter is added at the front of the phase-shifting circuit, which attenuates the average value of the equivalent acceleration of the detection noise by about 78 dB, and the average value is less than 0.1 μg, effectively reducing the impact of the detection noise on the accuracy of the QFA. In the following research work, the structure of the QFA will be optimized to reduce the value of the distributed capacitance, suppress the electric field coupling detection noise from the source, and improve the performance of the QFA.

Author Contributions: Conceptualization, Z.Z. and H.H.; methodology, Z.Z.; software, D.Z.; investigation, Z.Z., D.Z., H.H. and L.T.; writing—original draft preparation, Z.Z., D.Z. and Q.H.; writing—review and editing, Z.Z., D.Z., H.H., L.T. and Q.H. All authors have read and agreed to the published version of the manuscript.

Funding: This research was funded by the Open Research Fund of Hunan Provincial Key Laboratory of Flexible Electronic Materials Genome Engineering, grant number 202019, the Open Research Fund of the Hunan Province Higher Education Key Laboratory of Modeling and Monitoring on the Near-Earth Electromagnetic Environments, grant number N202107, and the Postgraduate Scientific Research Innovation Project of Hunan Province, grant number CX20200896.

Institutional Review Board Statement: Not applicable.

Informed Consent Statement: Not applicable.

Data Availability Statement: Not applicable.

Conflicts of Interest: The authors declare no conflict of interest.

References

1. Foote, S.A.; Grindeland, D.B. Model QA3000 Q-Flex accelerometer high performance test results. *IEEE Aerosp. Electron. Syst. Mag.* **1992**, *7*, 59–67. [CrossRef]
2. Beitia, J.; Clifford, A.; Fell, C.; Loisel, P. Quartz pendulous accelerometers for navigation and tactical grade systems. In Proceedings of the 2015 DGON Inertial Sensors and Systems Symposium (ISS2015), Karlsruhe, Germany, 22–23 September 2015; pp. 1–20.
3. Qin, H.; Yu, L. Review of Inertial Accelerometers at Home and Abroad. In Proceedings of the China Academy of Aerospace Electronics Technology Science and Technology Committee 2020 Annual Conference, Beijing, China, 15–16 December 2020; pp. 187–201.
4. Chang, G.; Zheng, B.; Zhao, Y.; Guang, N.; Zhang, Y. Demand analysis of accelerometer for air defense missile. In Proceedings of the The Third Chinese Conference on Autonomous Navigation (CCAN2019), Chongqing, China, 22–23 November 2019; pp. 1–7.
5. Guang, Y.; Wang, Y.; Zhao, P.; He, H.; Tang, L. Study of noise model for torquer in quartz flexible accelerometer header. *Transducer Microsyst. Technol.* **2018**, *37*, 34–37.
6. Huang, Y. *Analysis and Circuit Optimization of Distributed Capacitance in Differential Capacitance Measurements*; Changsha University of Science & Technology: Changsha, China, 2018.
7. Wang, X.; Song, L.; Ran, L.; Zhang, C.; Xiao, T. Drift Mechanism and Gain Stabilization of Digital Closed-Loop Quartz Flexible Accelerometer in Start-Up Performance. *IEEE Sens. J.* **2022**, *22*, 18649–18657. [CrossRef]
8. Cui, C.; Nie, L.; Li, L.; Wu, M.; Song, X. Improved design for temperature stability of quartz flexible accelerometer. *J. Chin. Inert. Technol.* **2022**, *30*, 658–665,673.
9. Zhang, C.; Wang, X.; Song, L.; Ran, L. Temperature Hysteresis Mechanism and Compensation of Quartz Flexible Accelerometer in Aerial Inertial Navigation System. *Sensors* **2021**, *21*, 294. [CrossRef] [PubMed]
10. Ran, L.; Zhang, C.; Song, L.; Lu, J. The Estimation and Compensation of the Loop-Parameter-Drifting in the Digital Close-Loop Quartz Flexible Accelerometers. *IEEE Access* **2020**, *8*, 5678–5687. [CrossRef]
11. Cao, J.; Wang, M.; Cai, S.; Zhang, K.; Cong, D.; Wu, M. Optimized Design of the SGA-WZ Strapdown Airborne Gravimeter Temperature Control System. *Sensors* **2015**, *15*, 29984–29996. [CrossRef] [PubMed]
12. Wang, Y.; Sun, X.; Huang, T.; Ye, L.; Song, K. Cold Starting Temperature Drift Modeling and Compensation of Micro-Accelerometer Based on High-Order Fourier Transform. *Micromachines* **2022**, *13*, 413. [CrossRef] [PubMed]
13. Wang, X.; Zhang, C.; Song, L.; Ran, L.; Xiao, T. Temperature compensation for quartz flexible accelerometer based on nonlinear auto-regressive improved model. *Meas. Control* **2022**, *56*, 124–132. [CrossRef]
14. Hou, W.; Liu, X.; Wu, W.; Wang, D. Output Noise Characteristics Analysis of the Quartz Flex Accelerometer. *Ship Electron. Eng.* **2015**, *35*, 39–42.
15. Wu, B.; Huang, T.; Ye, L.; Song, K. Noise shaping mechanism of rebalancing loop for inertial component. *J. Zhejiang Univ. (Eng. Sci.)* **2020**, *54*, 1819–1826.
16. Wen, Y.; He, H.; Tang, L.; Zhao, D.; Wang, Y.; Shang, M.; Liu, W.; Zeng, P. Read-out accuracy analysis and torquer-dirver design of quartz flexible accelerometer. *J. Mech. Electr. Eng.* **2014**, *31*, 1035–1039.
17. Huang, L.; Shao, Z.; Zheng, Y.; Li, L. Digital compensation for I/F converter of accelerometer. *J. Chin. Inert. Technol.* **2014**, *22*, 547–551.
18. Huang, Y.; Wang, S.; Liu, Q.; He, H. Noise model for accelerometer charging/discharging differential capacitance measurement circuit. *Transducer Microsyst. Technol.* **2019**, *38*, 53–55.
19. Ran, L.; Zhang, C.; Song, L.; Gao, S. Signal detection method of digital closed-loop quartz flexible accelerometer based on AC bridge. In Proceedings of the The Third Chinese Conference on Autonomous Navigation (CCAN2019), Chongqing, China, 22–23 November 2019; pp. 15–20.
20. Chen, B.; Song, L.; Zhang, C. Differential capacitance detection method based on capacitance bridge. *Transducer Microsyst. Technol.* **2019**, *38*, 116–119.
21. Zhang, Z.; Wang, Y.; He, H.; Tang, L. Analysis of Electric Field Coupling inside Quartz Flexible Accelerometer Header. In Proceedings of the 5th Chinese Conference on Autonomous Navigation (CCAN2021), Xiamen, China, 10–11 September 2021; pp. 77–84.

22. Beitia, J.; Loisel, P.; Fell, C. Miniature accelerometer for High-Dynamic, Precision Guided Systems. In Proceedings of the IEEE International Conference Symposium on Inertial Sensors and Systems, Kauai, HI, USA, 27–30 March 2017; pp. 35–38.
23. Lu, Y. Inertial Devices. In *Inertial Devices*; Aerospace Publishing House: Beijing, China, 1993; Volume 2, pp. 23–28.
24. Lotters, J.C.; Olthuis, W.; Veltink, P.H.; Bergveld, P. A sensitive differential capacitance to voltage converter for sensor applications. *IEEE Trans. Instrum. Meas.* **1999**, *48*, 89–96. [CrossRef]

micromachines

Article

A Derating-Sensitive Tantalum Polymer Capacitor's Failure Rate within a DC-DC eGaN-FET-Based PoL Converter Workbench Study

Dan Butnicu 

Basics of Electronics, Telecommunications and Information Technology Faculty, Technical University "Ghe.Asachi" of Iasi, 700050 Iasi, Romania; dbutnicu@etti.tuiasi.ro

Abstract: Many recent studies have revealed that PoL (Point of Load) converters' output capacitors are a paramount component from a reliability point of view. To receive the maximum degree of reliability in many applications, designers are often advised to derate this capacitor—as such, a careful comprehending of it is required to determine the converter's overall parameters. PoL converters are commonly found in many electronic systems. Their most important requirements are a stable output voltage with load current variation, good temperature stability, low output ripple voltage, and high efficiency and reliability. If the electronic system in question must be portable, a small footprint and volume are also important considerations—both of which have recently been well accomplished in eGaN transistor technologies. This paper provides details on how derating an output capacitor—specifically, a conductive tantalum polymer surface-mount chip, as this type of capacitor represented a step forward in miniaturization and reliability over previously existing wet electrolytic capacitors—used within a discrete eGaN-FET-based PoL buck converter determines the best performance and the highest MTBF. A setup based on an EPC eGaN FET transistor enclosed in a 9059/30 V evaluation board with a 12 V input voltage/1.2 V output voltage was tested in order to achieve the study's main scope. Typical electrical performance and reliability data are often provided for customers by manufacturers through technical papers; this kind of public data is often selected to show the capacitors in a favorable light—still, they provide much useful information. In this paper, the capacitor derating process was presented to give a basic overview of the reliability performance characteristics of tantalum polymer capacitor when used within a DC–DC buck converter's output filter. Performing calculations of the capacitor's failure rate based on taking a thermal scan of the capacitor's capsule surface temperature, the behavior of the PoL converter was evaluated.

Keywords: derating; DC–DC converter; eGaN; PoL; conductive polymer tantalum chip capacitor; reliability; MTBF

Citation: Butnicu, D. A Derating-Sensitive Tantalum Polymer Capacitor's Failure Rate within a DC-DC eGaN-FET-Based PoL Converter Workbench Study. *Micromachines* **2023**, *14*, 221. https://doi.org/10.3390/mi14010221

Academic Editor: Sadia Ameen

Received: 3 January 2023
Accepted: 13 January 2023
Published: 15 January 2023

1. Introduction

Power electronics designers know that, when it comes to reliability improvement, voltage is one of the strongest accelerators for the occurrence of failure mechanisms. As such, its lowering may significantly improve the MTBF component of DC–DC converters' output filters. The failure rate can be altered by the following parameters [1]:

- Voltage derating
- Temperature
- ESR (equivalent series resistance)

In this practical study, it is shown that a conductive tantalum polymer chip capacitor's MTBF is influenced by its voltage and voltage derating, which is a very effective way to increase the PoL converter's lifetime and reduce the capacitor's failure rate. Nevertheless, the reasons for the capacitor's voltage derating can have multiple causes depending on the capacitor technology and its applications. The equivalent series resistance (ESR) has become a kind of key metric for capacitors because it summarizes all the ohmic or resistive losses of the capacitor. Although tantalum capacitors provide low or good ESR performance, a new technology—conductive polymers—has been developed to provide a lower ESR value that exhibits conductivity characteristics close to those of metals. Compared to generic MnO_2-based tantalum capacitors, this new type—using conductive polymers—is significantly more conductive (100 times), and the direct benefits of this augmented conductivity are its lower ESR, improved high-frequency behavior, and superior reliability—especially for lower-rated voltages [2]. A conductive tantalum polymer chip capacitor is a capacitor with a solid electrolyte made of a conductive polymer. In the past decade, all polymer capacitor technologies seem to have been well adopted into the PoL power electronics converter branch. They have an anode body that is made of sintered tantalum powder, followed by the dielectric—which is a thin film of TaO (generated by electrochemical oxidation)—and, finally, the cathode, consisting of a highly conductive polymer layer deposited in the oxide layer [2–4]. As these Tantalum Polymer devices are actually polar capacitors, it is very important when using them in practice to give attention to the polarity marking. A reverse polarity will be allowed only up to the values indicated in the data sheet. A high voltage and high current stress will eventually lead to a rapid degradation of the components or device performance. Derating designs constitute a key factor in component selection for lowering the failure rate. When performing derating, it is necessary to keep the maximum stress at a proportion of the maximum ratings—this is called the derating factor [5–7]. If we act in voltage, derating means that the actual capacitor shall be used in the application at a lower voltage than the rated voltage.

When they are operated within their recommended guidelines, these Tantalum polymer chip devices—which are in fact solid state capacitors—demonstrate no wear out mechanism, as we can see in bathtub curve below. Although they can be operated at the full-rated voltage, professional circuit designers look for a minimum level of assurance in long-term reliability, which should be demonstrated with data. Since most applications do require long-term reliability, a voltage derating operation can provide the desired level of demonstrated reliability, based on industry-accepted acceleration models. Many manufacturers compulsorily recommend designers to consider voltage derating for the maximum steady-state voltage. Table 1 shows the recommendations for this issue used by one important producer of capacitors.

Table 1. Capacitor's derating for the range 2.5 V to 35 V Rated voltage; Above 105 °C, the voltage is de-rated linearly to 0.67 × Rated voltage up to the maximum operational temperature.

Rated Voltage		Derated Voltage	
−55 °C to 05 °C	**−55 °C to 125 °C**	**−55 °C to 105 °C**	**−55 °C to 125 °C**
2.5 V	1.7 V	2.3 V	1.5 V
4 V	2.7 V	3.6 V	2.4 V
6.3 V	4.2 V	5.7 V	3.8 V
10 V	6.7 V	9 V	6 V
16 V	10.7 V	12.8 V	8.6 V
20 V	13.4 V	16 V	10.7 V
25 V	16.8 V	20 V	13.4 V
35 V	23.5 V	28 V	18.8 V

In PoL and automotive applications—as described schematically in Figure 1—the output component position (capacitor C_{out}) selection follows Table 2, depending on the mission profile and the temperature requirements. For the 12 V line, the recommendation is—at minimum—a 35 V-rated voltage capacitor, taking into consideration the existing ISO Pulse requirements defined by the ISO7637 Specification.

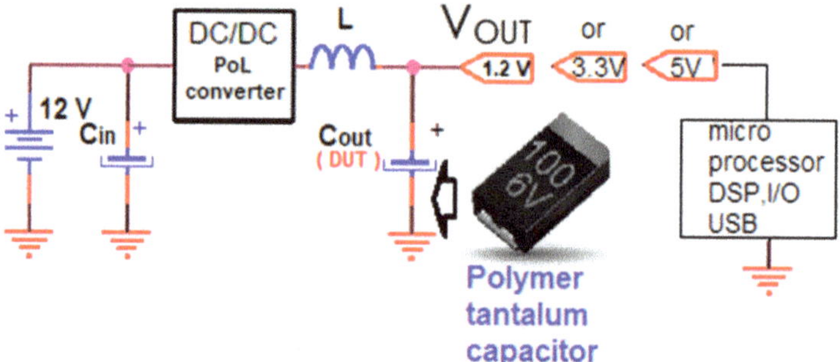

Figure 1. Schematic diagram for PoL and Automotive Applications.

Table 2. Capacitor's Application Voltage—Recommended Capacitors.

Application Voltage	Mission Profile Temperature (up to 105 °C)	Mission Profile Temperature (up to 125 °C)
<1 V	2.5 V	2.5 V
3.3 V	4 V	6.3 V
5 V	6.3 V	10 V
12 V	Minimum 35 V	Minimum 35 V

2. Materials and Methods

The operation of derating can be expressed by the percentage of the rated voltage that shall be subtracted. The aim of the derating is to reduce the number of stress factors applied to the capacitors. The two main stress factors are the voltage and temperature. The derating curve is shown in Figure 2, where V_R = the rated voltage, V_C = the rated voltage, T_{lc} = the lower category temperature, T_R = the rated temperature, and T_{uc} = the upper category temperature. Practically, if we carry out a 50% derating, this means that the capacitor shall be used at 50% of the rated voltage for a specific application (i.e., a 6.3 V-rated capacitor will be used on 3.15 V at maximum). Varying derating curves can be found in the MIL-HDBK-1547 military standard [8]. Due to new materials emerging on the market—corroborated by special manufacturing processes—polymer tantalum capacitors have a long way to go to meet automotive AEC-Q200 stress test requirements, and, consequently, require specific electrical transient tests too.

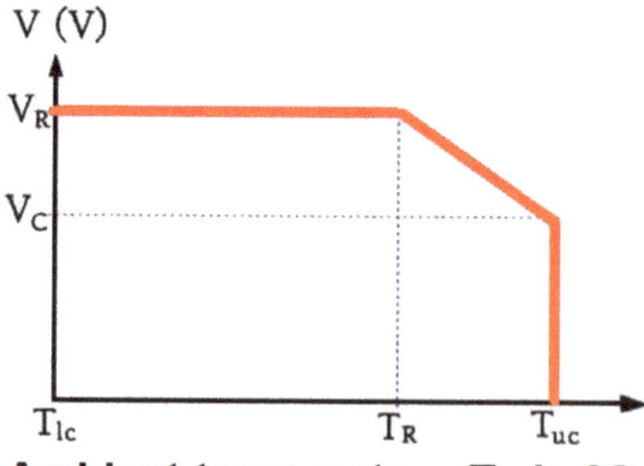

Figure 2. The derating curve.

For the experimental part of the study, an EPC9059/30V development board containing two EPC2100 eGaN (enhancement Gallium Nitrade) FET transistors in a half bridge configuration [9–12], using the Texas Instruments LM5113 gate driver, was used in a workbench hands-on investigation (Figure 3).

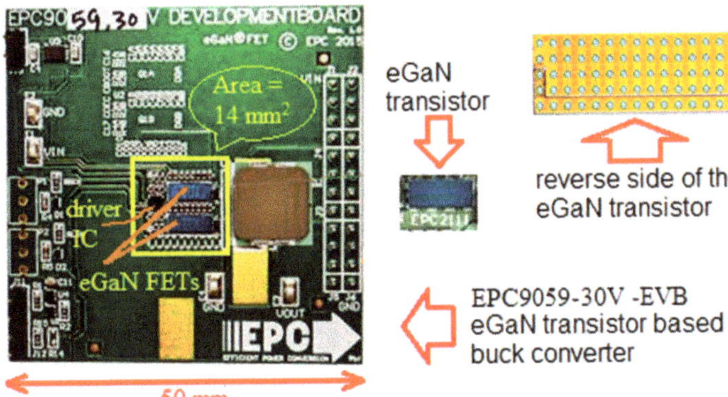

Figure 3. Planted area view of the eGaN evaluation board EPC9059-30V used in the experiment, with dimensions; the eGan-transistor's area is outlined in yellow and the reverse side of the eGaN transistors (EPC2111) has a Ball Grid Array (BGA packaging technology) [10].

The measurement connection diagram and the method for the temperature scanning are shown in Figure 4. Since the standard used for the reliability prediction required information about the capsule temperature of the component in question, an infrared thermal scanning device was used. The temperature collected by this was further used to calculate the piT stress factor. This, in turn, was used to calculate the lambda failure rate and then the MTBF. During the experiment, no forced cooling method was used for the investigated evaluation board, but only natural convection; also, the eGaN-FET transistors did not have a thermal radiator. Therefore, the extraction of information on the reliability of the converter was obtained in the most difficult conditions for thermal dissipation in active devices. The main parameters of the converter are shown in Table 3. For the DC–DC converter built with eGaN transistor technology, a tantalum polymer capacitor made with SMD technology-type encapsulation was selected for the output filter. Such converters are mission-critical in the telecom industry and computer applications, so knowledge of their capabilities in terms of their reliability is very important. This capacitor was the object of the voltage derating. In fact, two polymer capacitors—NEEgJ8, with a nominal voltage of 4 V and NEtjJ8, with a nominal voltage of 6.3 V—were used consecutively. Both were produced by NEC-Tokin Corporation headquartered in Shiroishi, Japan. (now part of KEMET). In Figure 5, these two capacitors and all the information related to their packaging, operating temperature, tolerance, identification of their technical parameters, and ESR (Equivalent Series Resistance) are shown. Additionally, on the right side of the figure, the actual values measured with a capacitor-tester for the electrical capacitance of each capacitor (255 µF vs. 220 µF, respectively) are shown, with which the calculations were made to find out the temperature stress factor π_T; then, this factor was entered into the formula for devising the failure rate.

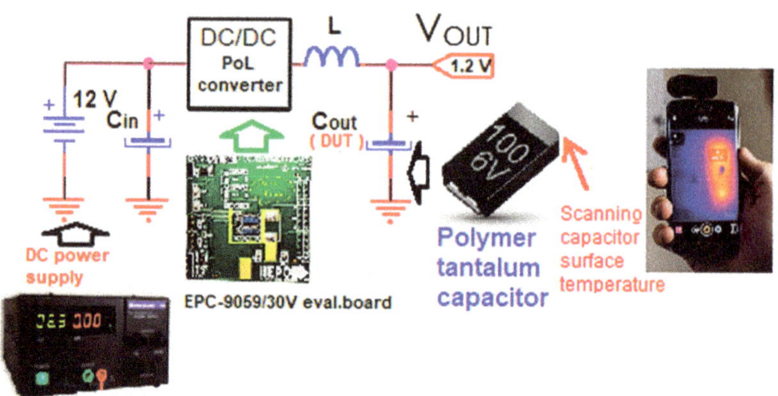

Figure 4. Diagram of connections with the derated polymer tantalum capacitors and how the capacitor's temperature is detected.

Table 3. Parameters for the investigated converter.

Parameter	Value
Load resistor	$R_{load} = 0.1\ \Omega$
Load current	$I_{out} = 12$ A
Input current	$I_{in} = 1.2$ A
Input voltage	$V_{in} = 12$ V
Output voltage	V_{out} =1.2V
Switching frequency	$f_{sw} = 300$ kHz
Inductor	$L = 3$ µH
Ambient temperature	~25 °C
Duty cycle	~13%
Output $C_{tantalum\text{-}poly\,A}$	220 µF/4 V
Output $C_{tantalum\text{-}poly\,B}$ [derated]	220 µF/6.3 V

Regarding the reliability calculation method, it must be said that the standard used provides a prediction and not an estimate. Reliability prediction for electronic components is commonly based on the failure rate, which is assumed to be constant during the lifetime period in the bathtub curve, as shown in Figure 6 [8,13].

$$R(t) = e^{-\lambda t} \quad (1)$$

where λ signifies the intrinsic failure rate, excluding early failures and wear-out failures. The mean time between failures (MTBF), i.e., when the reliability function has decayed to a value of e^{-1} = 36.7% (or in other words, only 37% of the units within a large group will last as long as the *MTBF* number) is:

$$MTBF = \lambda^{-1} \quad (2)$$

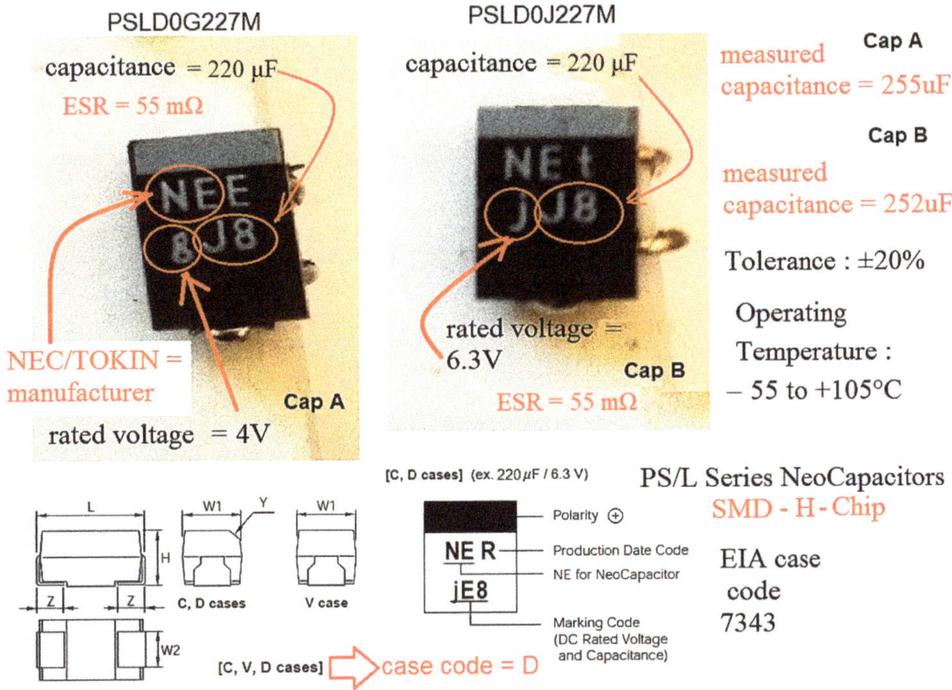

Figure 5. Packaging and identification information for the SMD polymer capacitors used for the reliability investigation of the eGan converter.

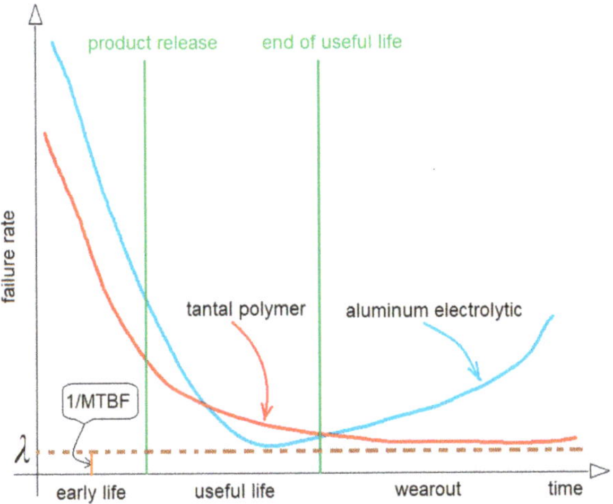

Figure 6. The bathtub curve.

The International Electrotechnical Commission released the IEC-TR-62380 standard in 2004, including features of newer components arriving on the market such as polymer capacitors, which did not appear in older standards such as MIL-HDBK-217 [8]. Thus, we performed the calculation of the capacitor's *MTBF* based upon the IEC-TR-62380 standard, in which a mathematical model for every electronic component is defined, in order to calculate their failure rates λ [13]. The reliability data taken from this prediction standard are taken mainly from field data concerning electronic equipment operating in that type of environment: *Ground; stationary; weather-protected* (means: equipment for stationary use on the ground in weather-protected locations, operating permanently or otherwise with controlled temperature and humidity and good maintenance). This applies mainly to telecommunications equipment and computer hardware. Experience has shown that component reliability is heavily influenced by mechanical and environmental conditions, as well as by electrical environment conditions. Estimated reliability calculations of equipment have to be carried out according to its field use conditions, so they are defined by the mission profile. A "mission profile" is usually defined as a table that includes all the details on the ambient temperature cycles during the lifetime of the device being discussed, on/off—state durations, numbers of operation cycles, etc. [8,13].

3. Results

3.1. Experimental Workbench SET-UP, Key Measurements, and the Mode of Investigation

An experimental workbench hands-on SET-UP was constructed in order to obtain a synchronous buck converter that lowers its voltage from 12 V to 1.2 V (a very often-used amount of voltage for powering FPGAs, microprocessors, etc.). The actual Capacitance, ESR, and V_{loss} (%) values for the two capacitors were measured using an BSIDE® *ESR02 Pro* transistor tester. The output filter reliability was improved by putting a higher-voltage rating capacitor on the output rail in parallel to the load resistance. Then, a thermal scan of the surface of the capacitor was executed in order to determine the temperature of operation, which is necessary in the calculation of the failure rate λ. In Figures 7 and 8 the thermal picture of the evaluation board area is shown, scanned by mean of a thermo-vision camera—specifically, the temperature of the capacitor's capsule was 36 °C for the 4 V-rated capacitor and 36 °C for the 6.3 V capacitator. These situations indicate to us that in the case of a derated capacitor, a smaller amount of current passes through it (i.e., a decreased power loss occurs). Later, using this information, the failure rate was calculated using the temperatures obtained and the models provided by the prediction standard chosen for the investigated capacitors. Finally, the MTBF was determined—a strong indicator of overall reliability. The scan process also showed that the highest temperature on the EVB is attributed to the coil, so this component contributed the most to decreases in the general reliability of the studied converter.

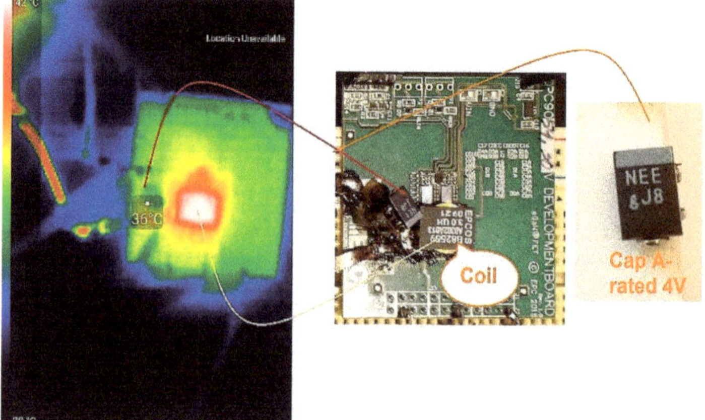

Figure 7. Thermal image of the Capacitor A surface. A 36 °C temperature is in evidence, as shown by the infrared thermo-vision device. Left—EVB scanned in infrared. Middle—EVB seen from the side, plated with components. Right—Photo of the investigated capacitor.

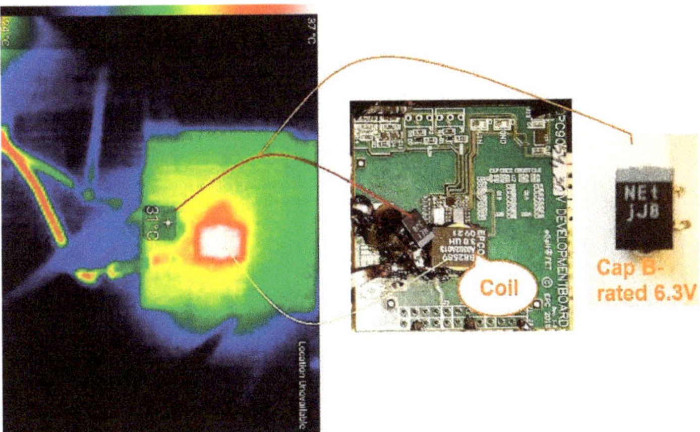

Figure 8. Thermal image of the Capacitor B surface (derated). A 31 °C temperature is in evidence, as shown by the infrared thermo-vision device. Left—EVB scanned in infrared. Middle—EVB seen from the side, plated with components. Right—Photo of the investigated capacitor.

This is in agreement with the Arrhenius law (Figure 9), which stipulates that if the capacitor's temperature increases by 10 °C, then the reliability will decrease by half.

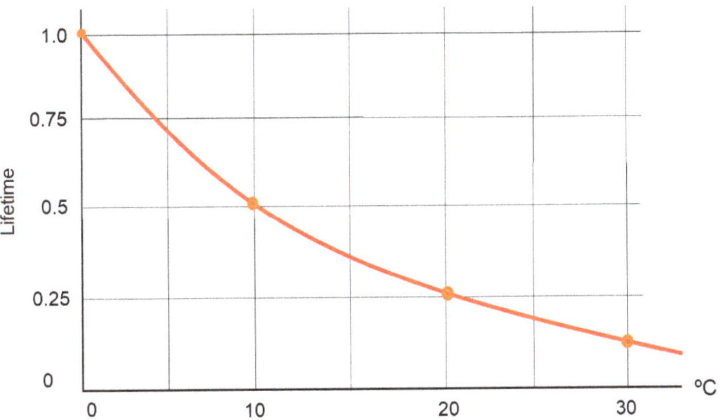

Figure 9. Arrhenius Law.

The good operation of the converter was experimentally validated by measuring some important waveforms, such as:

- The intensity of the current through the converter's inductance L = 3 μH (a SMD power coil constructed with flat wire windings), which was collected by a current probe. The current probe used in the experiment had a 10-mV output voltage, corresponding to 4 A as measured.
- The voltage measured in the so-called "switching point" (at the junction between high FET and synchro FET).

The most important waveforms measured on the converter under load at room temperature are shown in Figure 10. These waveforms ensure the CCM mode's functioning in the converter—i.e., Continuous Conduction Mode. A general view of the experimental set-up is shown in Figure 11.

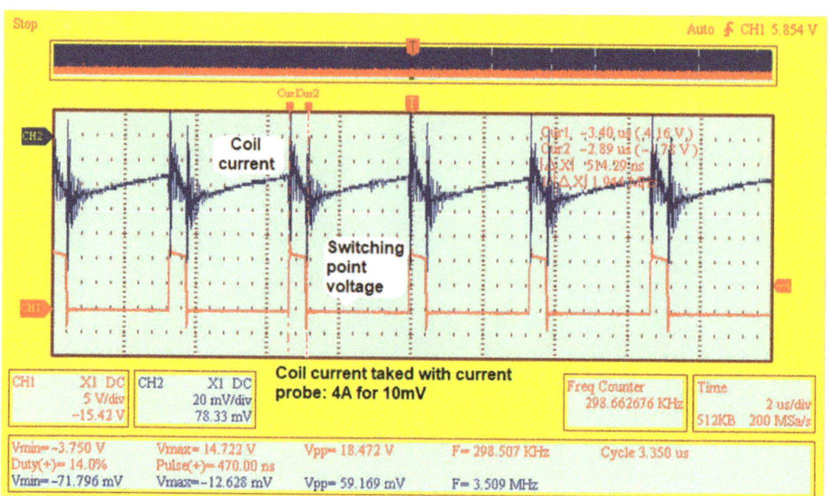

Figure 10. Basic waveforms of the PoL DC–DC eGaN-FET-based converter, ensuring the CCM mode of operation.

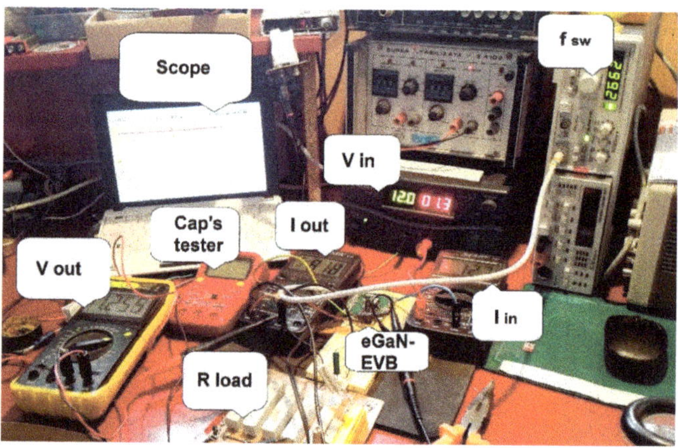

Figure 11. General view of the experimental set-up. Voltage waveforms were displayed using an USB-DSO oscilloscope—ISDS 220B and a CP-05$^+$ current probe.

3.2. Reliability Calculation

In order to execute the calculation for the tantalum polymer capacitor, the mathematical model for this type of capacitor was taken from the IEC-TR-26380 standard [13], as shown in Figure 12.

$$\lambda = 0.4 \times \{[(\Sigma(\pi_T)_i \times \tau_i)/(\tau_{on} + \tau_{off})] + 3.8 \times 10^{-3} \times [\Sigma(\pi_n)i \times (\Delta T_i)^{0.68}]\} \times 10^{-9}/\text{h} \quad (3)$$

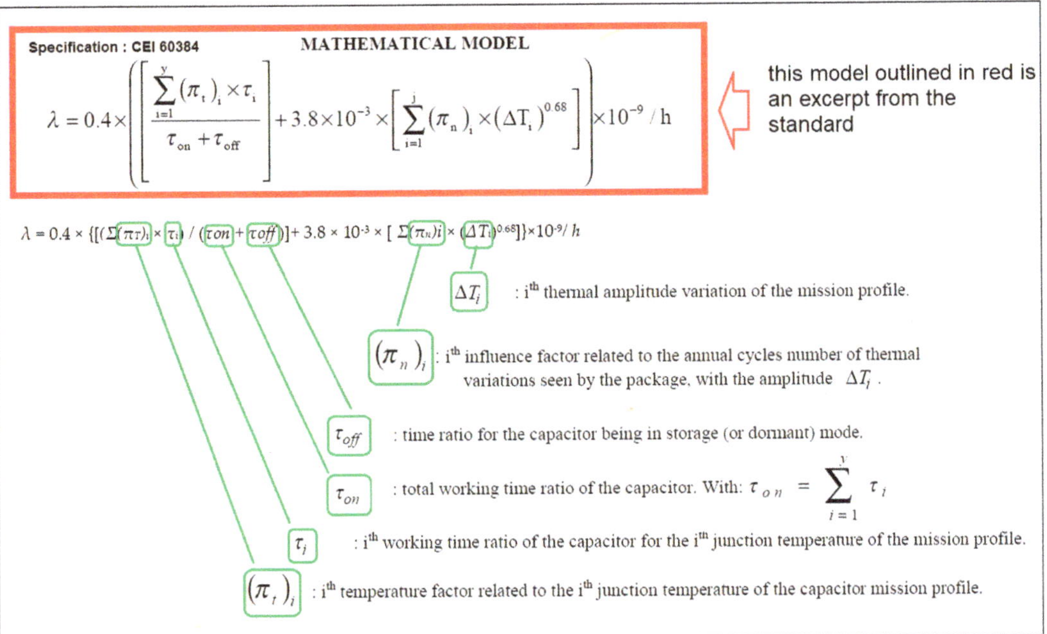

Figure 12. Explanation of how the failure rate λ is devised using the models for Tantalum Polymer Capacitors (the models are provided by the IEC-TR-62380 Standard). The mathematical model and parameter's formulae, outlined in green, are depicted as they appear in this standard.

The Standard specifies that the above formula gives field values if the ratio peak voltage/rated voltage is less than or equal to 0.8 (with peak voltage = continuous voltage + peak value of the alternative voltage). In the investigated circuit, the peak voltage was 1.5 V and the maximum rated voltage for the capacitors was 6.3 V, so the ratio was equal to 0.238 < 0.8.

After performing some preliminary calculations, which we will not reproduce here as they are irrelevant, it was determined that:

- τ_i = 365 days × duty cycle = 365 × 0.13
- π_T = 0.9 derived from Figure 11 @ 25 °C
- n_i = 365 and $(\pi_n)_i = (n_i^{0.76}) \times 1.7$
- $\tau_{on} = 1$ and $\tau_{off} = 0$
- $\Sigma(\pi_n)_i = n_i$ and $\Sigma(\pi_T)_i = \pi_T$
- ΔT_i = capacitor's surface temperature—ambient temperature

where ΔT_i is derived using the actual temperature of the capacitor's capsule and π_T from Figure 13.

Figure 13 shows the temperature factor at an ambient temperature of ~25 °C, at which the experimental measurements took place. Many details are given in subchapter 8, mission profiles, within the IEC-TR-62380 standard [13]—hence:

- $\lambda_{polimer\ capA} = 0.4 \times \{[(0.9 \times 54.75)/(1 + 0)] + 1.4 \times (10^{-3}) \times [(1.7 \times 365^{0.6}) \times (36 - 25)^{0.68}]\} \times (10^{-9})/\mathrm{h}$ = 7215.425 FIT = 7.215425 F/10^6 h
- $\lambda_{polimer\ capB}$[derated] $= 0.4 \times \{[(0.9 \times 54.75)/(1 + 0)] + 1.4 \times (10^{-3}) \times [(1.7 \times 365^{0.6}) \times (31 - 25)^{0.68}]\} \times (10^{-9})/\mathrm{h}$ = 6821.755 FIT = 6.821755 F/10^6 h

where FIT stands for Failures in time (1 FIT = one failure in 10^9 h) and F/10^6 h stands for Failures in one million hours.

Consequently, we have:

- $MTBF_{polimer\ capA}$ = 138,591 h or 15.821 yr
- $MTBF_{polimer\ capB}$[derated] = 146,589 h or 16.734 yr

The failure rate in FIT (failures per 10^9 h) and the MTBF in hours for the two investigated capacitors are presented in a suggestive comparative diagram, shown in Figure 14.

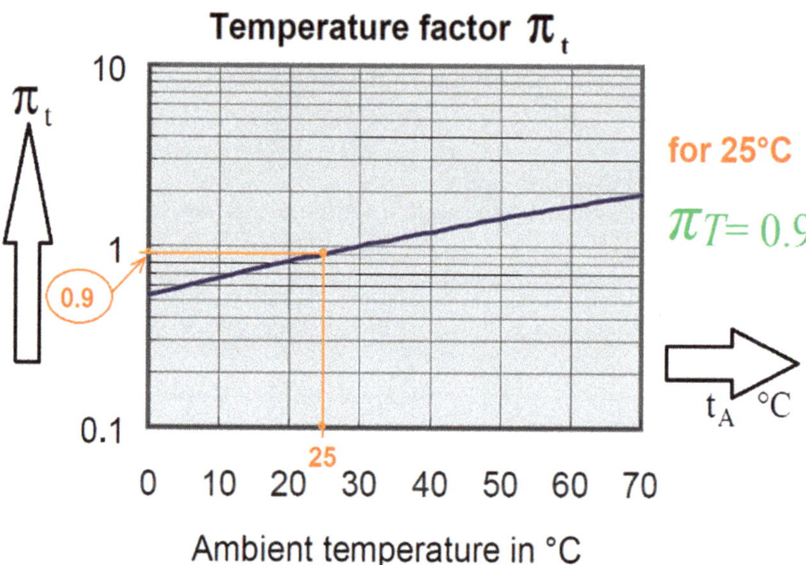

Figure 13. Temperature factor π_T versus ambient temperature for the tantalum polymer capacitor (graphical derived—orange lines—using a nomogram from the IEC-TR-62380 standard at 25 °C [13]).

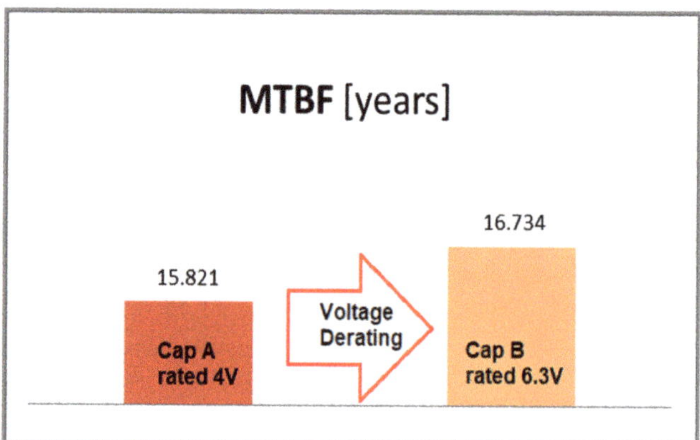

Figure 14. Graphical comparison of the MTBF for the two capacitors.

4. Discussion

Derating the stress levels of electronic components is recognized as a very effective way of improving the device's reliability and also for avoiding failures due to overstress. This paper provides a practical study that refers to a simple way of augmenting the reliability of the output filter of modern eGan transistor-based DC–DC converters. Experimental results combined with calculations based on data and models provide by the IEC-TR-62380 reliability prediction standard were validated by performing a voltage derating operation upon the converter's output capacitor, resulting in an improvement in the MTBF of about

7%. This was possible using a 6.3 V-rated polymer tantalum capacitor instead of a 4 V-rated capacitator on a 1.2 V-output voltage rail.

Funding: This paper was financially supported by the Project "Network of excellence in applied research and innovation for doctoral and postdoctoral programs/InoHubDoc"; the project was co-funded by the European Social Fund financing agreement no. POCU/993/6/13/153437. This project takes place at Technical University "Ghe.Asachi" of Iasi, Iasi 700050, Romania.

Data Availability Statement: Not applicable.

Acknowledgments: The author would like to thank to Andrei Alistar from Continental Automotive, Iaşi for providing the EPC 9059/30V eGaN-FET evaluation board.

Conflicts of Interest: The author declares no conflict of interest.

References

1. Anon. Voltage Derating Rules for Solid Tantalum and Niobium Capacitors. Available online: https://www.avx.com/docs/techinfo/VoltageDeratingRulesforSolidTantalumandNiobiumCapacitors.pdf (accessed on 14 January 2023).
2. Anon. Available online: chrome-extension://efaidnbmnnnibpcajpcglclefindmkaj/https://nepp.nasa.gov/docuploads/0EA22600-8AEC-4F47-9FE49BAABEAB569C/Tantalum%20Polymer%20Capacitors%20FY05%20Final%20Report.pdf (accessed on 14 January 2023).
3. Anon. Polymer Tantalum Capacitors Continue to Toughen Up to Meet Automotive Demands. Available online: https://www.kemet.com/en/us/technical-resources/polymer-tantalum-capacitors-continue-to-toughen-up-to-meet-automotive-demands.html (accessed on 14 January 2023).
4. Freeman, Y. *Tantalum and Niobium-Based Capacitors*, 1st ed.; Springer: Berlin/Heidelberg, Germany, 2017.
5. Freeman, Y.; Lessner, P. Evolution of Polymer Tantalum Capacitors. *Appl. Sci.* **2021**, *11*, 5514. [CrossRef]
6. Ma, C.; Wang, K.; Wang, J.F. Control Design and Realization of a Fast Transient Response, High-Reliability DC-DC Converter With a Secondary-Side Control Circuit. *IEEE Access* **2021**, *9*, 36976–36985. [CrossRef]
7. Caldwell, B.; Bechtold, L.; Douty, B.; Murray, J. Optimized Derating to Support Reliable System Design. In Proceedings of the 2020 Annual Reliability and Maintainability Symposium (RAMS), Palm Springs, CA, USA, 27–30 January 2020; pp. 1–5. [CrossRef]
8. Anon. Available online: http://everyspec.com/MIL-HDBK/MIL-HDBK-1500-1799/MIL_HDBK_1547A_2102 (accessed on 14 January 2023).
9. Reusch, D.; Strydom, J. Evaluation of Gallium Nitride Transistors in High-Frequency Resonant and Soft-Switching DC-DC Converters. *IEEE Trans. Power Electron.* **2014**, *30*, 5151–5158. [CrossRef]
10. Anon. Available online: https://epc-co.com/epc/products/demo-boards/epc9059 (accessed on 14 January 2023).
11. Anon. Available online: https://epc-co.com/epc/products/gan-fets-and-ics/epc2100 (accessed on 14 January 2023).
12. Anon. Available online: chrome-extension://efaidnbmnnnibpcajpcglclefindmkaj/https://www.ti.com/lit/ds/symlink/lm5113.pdf (accessed on 14 January 2023).
13. IEC. *TR 62380: Reliability Data Handbook*; IEC: Geneva, Switzerland, 2006.

MDPI
St. Alban-Anlage 66
4052 Basel
Switzerland
www.mdpi.com

Micromachines Editorial Office
E-mail: micromachines@mdpi.com
www.mdpi.com/journal/micromachines

www.ingramcontent.com/pod-product-compliance
Lightning Source LLC
LaVergne TN
LVHW070659170726
843515LV00004B/445
9783036595115